I0050925

Working with Fugacity

Multimedia environmental models are powerful tools for predicting the fate and distribution of chemicals in the environment. They are used for assessing the potential exposure of ecosystems and humans to contaminants, as well as for developing and evaluating chemical management approaches. This book has been developed with the aim of providing support to students and researchers who are new to fugacity-based environmental modelling or wish to refresh their knowledge and skills. The nature and approach of fugacity-based calculations are developed methodically and sequentially from the very first steps, with computational details explicitly shown and discussed. Whereas our text *Multimedia Environmental Models: The Fugacity Approach* gives detailed justification for the theoretical underpinnings of the fugacity approach, this workbook instead focusses on the step-by-step exposition of computational details and approaches without excessive theoretical justification. We believe that this approach will serve to help clarify the necessary computational steps and approaches needed to design and implement fugacity-based computations in real-life environmental contexts.

Features:

- Step-by-step development of computations and models
- Sequenced introduction of topics to help develop computational facility
- Screen shots of model and worksheet calculations with equations revealed for clarity
- Explicitly detailed calculations and methodology
- Many examples chosen from published works of realistic applications

Working with Fugacity

A Multimedia Environmental Modelling Workbook

J. Mark Parnis and Donald Mackay

CRC Press
Taylor & Francis Group
Boca Raton London New York

CRC Press is an imprint of the
Taylor & Francis Group, an **informa** business

First edition published 2026
by CRC Press
2385 NW Executive Center Drive, Suite 320, Boca Raton FL 33431

and by CRC Press
4 Park Square, Milton Park, Abingdon, Oxon, OX14 4RN

CRC Press is an imprint of Taylor & Francis Group, LLC

Library of Congress Cataloging-in-Publication Data
Names: Parnis, J. Mark author | Mackay, Donald, 1936- author
Title: Working with fugacity : a multimedia environmental modelling
workbook / J. Mark Parnis and Donald Mackay.
Description: First edition. | Boca Raton FL : CRC Press, 2026. | Includes
bibliographical references. | Summary: "Multimedia environmental models
are powerful tools for predicting the fate and distribution of chemicals
in the environmental. They are used for assessing the potential exposure
of ecosystems and humans to contaminants, as well as for developing and
evaluating chemical management approaches. This workbook has been
developed with the aim of providing support to students and researchers
who are new to fugacity-based environmental modelling or wish to refresh
their knowledge and skills"-- Provided by publisher.
Identifiers: LCCN 2025039669 (print) | LCCN 2025039670 (ebook) |
ISBN 9781041108665 hardback | ISBN 9781041108658 paperback |
ISBN 9781003657170 ebook
Subjects: LCSH: Environmental chemistry--Mathematical models |
Thermodynamics--Mathematical models
Classification: LCC TD193 .P37 2025 (print) | LCC TD193 (ebook)
LC record available at https://lccn.loc.gov/2025039669
LC ebook record available at https://lccn.loc.gov/2025039670

ISBN: 978-1-041-10866-5 (hbk)
ISBN: 978-1-041-10865-8 (pbk)
ISBN: 978-1-003-65717-0 (ebk)

DOI: 10.1201/9781003657170

Typeset in Times
by KnowledgeWorks Global Ltd.

Contents

Preface

Many years ago, I had the pleasure of hearing Don Mackay speak at Trent University on the subject of environmental chemical fate modelling, a discipline that he helped to pioneer over the last 50 years at The University of Toronto and at Trent University. At that time, I was a young chemistry professor, busy setting up a new low-temperature spectroscopy lab. We spoke after the talk, and at that time he mentioned the need for computational techniques for estimating physico-chemical properties of chemicals that had not yet been characterized experimentally. I filed this interesting idea away for a couple of decades, and then about 12 years ago, I approached Don to discuss following up on this idea. He was keen and enthusiastic, and we began a fruitful and exciting collaboration that included application of COSMO-RS theory to estimation of physico-chemical properties of environmentally relevant chemicals, polymeric passive sampling media, indoor dust, and other materials, thereby realizing his hopes for such computation-based techniques to be available and applied in environmental chemistry.

During this collaborative period, Don mentioned his desire to write a "Workbook" on fugacity applications in environmental modelling. His hope was to write a book that was heavy on computational detail and light on theoretical justification, something that a student could use in which no steps were skipped or skimmed. I initially set out to draft such a workbook based on the 2nd edition of Don's "Multimedia Environmental Modelling: The Fugacity Approach." That attempt was a failure; instead it generated the third edition of the text, with additional materials on newer estimation methods, new graphics and typesetting and elimination of some materials that were now available widely in textbooks in the field.

After the third edition came out, I once again set out to try to draft the Workbook, and the book in your hands is the result of this effort. During its development, Don was in increasing physical decline but was able to read and edit all chapters and provide direction and insights based on his long career's experience and wisdom. Sadly, Don did not live to see this book project fully realized, but his vision and insights live on in this and all his published works. It is with great respect and gratitude to Don that this work is brought to you, in hopes that it will contribute to realizing his vision of helping students to recognise the great power and promise of the fugacity approach in multimedia modelling.

J. Mark Parnis
Peterborough, ON Canada
April 8, 2025

About the Authors

J. Mark Parnis is an Emeritus Professor of Chemistry at Trent University in Peterborough, Ontario, Canada. A graduate of the University of Toronto, he is a physical chemist with early research work in metal atom and cluster reactions with organic molecules and a teaching emphasis on quantum mechanics, spectroscopy, and kinetics. In more recent years, he joined forces with Donald Mackay to focus on the development and implementation of techniques for estimating physico-chemical properties of molecular species with an emphasis on environmental modelling applications. He is a former director of the Canadian Environmental Modelling Centre at Trent University and continues to work on applications of fugacity-based models and property estimation techniques in environmental modelling. He maintains and updates the various models developed by Don Mackay's group, which are available from the CEMC Website (https://www.trentu.ca/cemc/resources-and-models).

Donald Mackay was an internationally renowned engineer and scientist, the acknowledged pioneer of fugacity-based modelling applications in environmental fate and exposure methodology. Don graduated from the University of Glasgow and was most recently an Emeritus Professor in the School of the Environment at Trent University. He was also Professor Emeritus in the Department of Chemical Engineering and Applied Chemistry of the University of Toronto where he taught for some 30 years and established himself as a pioneer of multimedia modelling in environmental science. Moving to Trent University in 1995, he contributed to the growth and maturation of the environmental science program and established the Canadian Environmental Modelling Centre, which he led until his official retirement in 2002. Since that time, Don continued to work in the field, producing over 750 articles, many books, and numerous reports during his career. The recipient of the Order of Canada and many other awards, Don was a leading figure in the field of environmental fate modelling.

Introduction

Fugacity-based chemical fate modelling is a fascinating and sometimes complicated scientific pursuit. Such modelling is of critical importance in the early stages of chemical management, when new chemicals are being introduced into production and little is known yet about their actual behaviour and fate in the environment. As a tool for rapid screening of persistence, bioavailability, and transport, it is unmatched in its power to predict these and other related key chemical management properties. In spite of the fact that such fugacity-based models are easy to use, they are not as easy to develop. Their conception and practical realisation require a detailed knowledge of the principles of equilibrium chemical partitioning as well as steady-state and dynamic transport and diffusion processes. Our text "Multimedia Environmental Models: The Fugacity Approach" details the underlying theory and approach to developing such models. Despite the fact that this text introduces and develops a number of fundamental modelling scenarios, there remains a need for a bridge between the theoretical underpinnings of fugacity-based tools, and their hands-on application in various environmental settings of current interest. It is with this in mind that we have developed this book.

The aim of this workbook is to give fugacity modelling learners access to highly detailed examples of fugacity-based calculations for use in models of chemical fate in the environment. Each example system has been worked out in a systematic and detailed manner, with supporting commentary that is normally omitted from published works. For this reason, there is a degree of repetition that is not normally found in textbooks, to help reinforce key principles and approaches. We feel that this is important to develop a ready familiarity with the material, and to cultivate an organized and logical approach to developing and realizing useful chemical fate models. Once these systems have been mastered, the reader should be well-equipped to deal with other chemical fate modelling scenarios with confidence.

Many equations and relations are presented in this workbook, with key equations shown in a highlighting rectangle. In some cases, the theoretical underpinnings are given, but in other cases, the practical "bottom-line" equation or relation is presented with minimal or no derivation. This is intentional, as the aim of this book is to help develop fluency with the practical aspects of multimedia fugacity-based modelling, rather than to provide a full theoretical foundation to all aspects of the calculations. As well, we have chosen to use referencing only sparingly, since virtually all of the necessary derivations and their supporting referenced work is given in "Multimedia Environmental Models: The Fugacity Approach".

We have developed a somewhat more detailed symbol system for this book, compared with the third edition of "Multimedia Environmental Models: The Fugacity Approach". Wherever possible, we have adhered to a systematic use of subscripts for media designations, and superscripts for compartment designation, process designations and other such notations. Notation from the 3rd edition was preserved wherever possible, but when greater clarity could be achieved, we chose to modify the symbol accordingly. It is hoped that any frustration with this modified symbol system will be rewarded with greater transparency for learners first encountering this field of modelling.

Multimedia environmental fate models are comprised of dozens of chained calculations which suffer from rounding errors if care is not taken along the way. Throughout the workbook, we have also taken care to avoid rounding errors in sequences of connected calculations by carrying one insignificant digit as a subscripted value. At any point in the calculation, these may be rounded off to get the appropriate "final" answer that would apply if the calculation were to end there. In this book, rounding is only done for the final answer. Note that in questions where whole numbers are given, each digit is considered to be significant, including zeros. For example, 1000 L would be treated as 1000. in the calculation.

Overview of the Fugacity-Based Approach

Fugacity is a thermodynamically defined "escape tendency" of a chemical from a given environment. A key environmental modelling application of fugacity is as an equilibrium criterion, wherein any two or more compartments in contact at equilibrium will have the same fugacity *f*, even though their concentrations are not the same:

$$\boxed{f_A^{Eq} = f_B^{Eq} = f_C^{Eq} = \ldots}$$

$$\boxed{C_A^{Eq} \neq C_B^{Eq} \neq C_C^{Eq} \neq \ldots}$$

The fundamental relationship between concentration *C* and fugacity is:

$$\boxed{C_i = Z_i f_i}$$

Here, the constant of proportionality is the fugacity capacity or *Z*-value. The greater the fugacity capacity of a medium for a chemical, the higher that chemical's concentration has to be in that medium to achieve a given fugacity. Fugacity capacities for various media are summarized in Table 0.1 at the end of this section, as used in this workbook and the work of Mackay and associates.

Fugacity capacities are related to equilibrium partition ratios *K* as follows, and as such they may be viewed as "half a partition ratio":

$$\boxed{K_{AB} = \frac{Z_A}{Z_B}}$$

Note that, whereas the partition ratio is a property of a given chemical equilibrated between two media compartments, the fugacity capacity is a property of that chemical in only one media compartment and is therefore "transferrable" to various partition ratios that might involve that medium.

For a medium that is composed of more than one subcomponent (i.e., is modelled with more than one compartment), the medium may be treated as a homogeneous entity by defining a bulk fugacity capacity, which is the volume fraction $\left(v_i^f\right)$ weighted sum of the individual subcomponent fugacity capacities. Therefore, in general:

$$\boxed{Z_i^{Bulk} = v_i^{f-A} Z_A + v_i^{f-B} Z_B + v_i^{f-C} Z_C + \ldots}$$

All processes by which a chemical may move between any two compartments are represented by a *D*-value which is similar to a rate constant. For bulk or advective transport into and out of a compartment, the rate of transfer *r* is given by:

$$\boxed{r_A^{Adv} = f_A D_A^{Adv}}$$

Here, the D-value for advection is related to the medium flow rate G and concentration as follows:

$$\boxed{G_iC_i = (G_iZ_i)f_i = D_i^{Adv} f_i} \text{ where } \boxed{D_i^{Adv} = G_iZ_i}$$

For any degradation process that removes or transforms the chemical, we assume a first-order loss and relate the rate constant k to the D-values for transformation as:

$$\boxed{k_iV_iC_i = (k_iV_iZ_i)f_i = D_i^{Deg} f_i} \text{ where } \boxed{D_i^{Deg} = k_iV_iZ_i}$$

Here, V_i is the volume of the medium in question.

Diffusive transport between two compartments that are not at equilibrium occurs at a rate that is proportional to the difference in fugacity between the compartments:

$$\boxed{r_{AB}^{Diff-Ov} = D_{AB}^{Diff-Ov}(f_A - f_B)}$$

The Whitman Two-Resistance theory approach defines the overall D-value for diffusive transport between two media compartments as:

$$\boxed{D_{AB}^{Diff-Ov} = \left(\frac{1}{\left(\frac{1}{D_{AB}^{Diff-A}} + \frac{1}{D_{AB}^{Diff-B}}\right)}\right)}$$

When systems are not at steady state, modelling must be done with time-dependence expressed in terms of differential equations. In these cases, such equations are constructed using a rate-balance approach for each compartment, based on relating the net rate of change in fugacity to the rates of input and output to the compartment:

$$\frac{df_i}{dt} = \left(\frac{1}{V_iZ_i}\right)\left[I_i + \sum_{j \neq i}^{n}(D_{ij}f_j) - D_i^T f_i\right]$$

where I_i is the chemical input rate for the compartment in question "i", D_{ij} is the D-value for any transport process such as inflow advection or intermedium diffusion from compartment "j", and D_i^T is the sum of all parallel loss processes that remove the chemical from compartment "i". The latter may include outflow advection, transformation losses and diffusion out of compartment "i" to one or more compartments with which it is in contact. As there will be as many such differential equations as there are compartments, and each compartment has its own unknown fugacity, there are enough equations to solve for the fugacities.

For any of the above systems, the amount of chemical in a compartment is given by:

$$\boxed{m_i = V_iZ_if_i}$$

Here, V_i is the volume of the compartment in question.

The molar rate for any process is given by:

$$\boxed{r_i = D_if_i}$$

TABLE 0.1
Fugacity capacities (*Z*-values) for some common media

Compartment	Fugacity Capacity (*mol Pa*$^{-1}$ *m*$^{-3}$)	Notes
Air	$Z_A = \frac{1}{RT}$	R = gas constant ($m^3\ Pa\ K^{-1}\ mol^{-1}$) T = Kelvin temperature (K)
Water	$Z_W = \frac{1}{H} = \frac{C_W^{Sat}}{P^{Sat}}$	H = Henry's Law ($Pa\ m^3\ mol^{-1}$) constant for chemical C_W^{Sat} = Saturation concentration of chemical in water ($mol\ m^{-3}$) P^{Sat} = Partial pressure of the chemical (in liquid or sub-cooled state) above saturated solution in water (Pa)
Octanol	$Z_O = K_{OW} Z_W$	K_{OW} = octanol-water partition ratio
Soil, sediment	$Z_{Medium} = m_{Medium}^{f-OC} \times 0.35\,L\ kg^{-1} \times K_{OW} \times Z_W \times \left(\frac{\rho_{Medium}\left(kg\,m^{-3}\right)}{1000\,L\,m^{-3}} \right)$	m_{Medium}^{f-OC} = mass-fraction of organic carbon in medium ρ_{Medium} = density of medium ($kg\ m^{-3}$)
Biota	$Z_{Biota} = v_{Lipid}^{f} Z_O = v_{Lipid}^{f} K_{OW} Z_W$	v_{Lipid}^{f} = volume fraction of lipid in biota

1 Equilibrium Partitioning in Closed Systems

All about Z-Values

Equilibrium partitioning involves using the equifugacity condition to determine the concentrations and molar or mass amounts of a chemical released into a model environment of two or more compartments that are at equilibrium. In the first step, the calculations involve determining the fugacity capacity of all compartments (*Z*-values) and then using mass balance to determine the overall system fugacity. With that in hand, the compartment concentrations and amounts can be determined directly. Partition ratios are used to determine some *Z*-values from others, usually starting from the fugacity capacity of air, then of water using the Henry's Law constant, and then all others using the octanol–water partition ratio and lipid or organic carbon fractions.

1.1 CONCENTRATION-BASED CALCULATIONS

The simplest environmental modelling scenario possible is one in which a chemical is released into a single environmental compartment "*i*" and it remains there unchanged indefinitely. In this case, a trivial calculation of concentration in that compartment C_i applies, based on the molar quantity m_i of chemical released and the volume of the compartment V_i:

$$\boxed{C_i = \frac{m_i}{V_i}}$$

Note that in this book we use "*m*" to represent the overall amount of the chemical in moles, as opposed to the more traditional use of "*n*" for this quantity. In the present work, "*n*" is generally reserved for the number of phases in a system.

Worked Example 1.1

30.0 *kg* of a chemical with molar mass, $M = 200.0\ g\ mol^{-1}$ is added to a tailing pond of volume, $V = 4.00 \times 10^5\ m^3$. Calculate the molar concentration of the chemical in the pond in $mol\ m^{-3}$ units, assuming complete mixing.

First, calculate the total number of moles of the chemical *m* as the mass of chemical in *kg* divided by the molar mass, *M*, in $g\ mol^{-1}$, with the necessary mass unit conversion from kilogram to gram:

$$m = \frac{mass}{M}$$

$$m = \frac{30.0\,kg \times 10^3\,g\,kg^{-1}}{200.0\,g\,mol^{-1}} = 1.50_0 \times 10^2\,mol$$

Now calculate the molar concentration in the water in $mol\ m^{-3}$ units:

$$C_W = \frac{m}{V_W} = \frac{1.50_0 \times 10^2\,mol}{4.00 \times 10^5\,m^3} = 3.75_0 \times 10^{-4}\,mol\,m^{-3}$$

Note that we express molar concentration in moles per m^3 throughout.

DOI: 10.1201/9781003657170-1

Word of Warning! Chemists are used to expressing concentrations of chemicals in units of moles per litre of solution. In fugacity-based work concentration of a chemical in a medium is expressed in units of moles per cubic metre of medium.

Here, we assume that the volume of the environmental compartment is much greater than the volume of chemical added, such that the volume of the chemical (and its solvent carrier, if appropriate) can be ignored.

For purposes of constructing mass balance expressions, we can express the molar amount of chemical in a given compartment "i" in terms of the compartment concentration and volume:

$$\boxed{m_i = C_i V_i}$$

Worked Example 1.2

Calculate the number of moles of chemical in the tailing pond from the previous example, based on the concentration determined there.

The calculation is a simple substitution of molar concentration and compartment volume into the expression for the number of moles:

$$m_W = C_W V_W = 3.75_0 \times 10^{-4}\, mol\, m^{-3} \times 4.00 \times 10^5\, m^3 = 1.50 \times 10^2\, mol$$

If one compartment is in contact with another for long enough to establish an equilibrium distribution of the chemical between the two compartments, we may express the relationship between the concentrations in each compartment in terms of the equilibrium constant for partitioning, known as the partition ratio K_{AB}, where A and B are the names or symbols for the two equilibrating compartments:

$$\boxed{K_{AB} = \frac{C_A}{C_B}}$$

Figure 1.1 illustrates this relationship. Here, the naming of the partition ratio is in keeping with the sense of the fraction, in that the first-named compartment in the subscript corresponds to the concentration term in the numerator and the second-named compartment corresponds to the denominator. Note that, historically, the partition ratio has been generally referred to as a "partition coefficient". However, the International Union of Pure and Applied Chemistry (IUPAC) recommends the use of the term "partition ratio", which we adopt here.

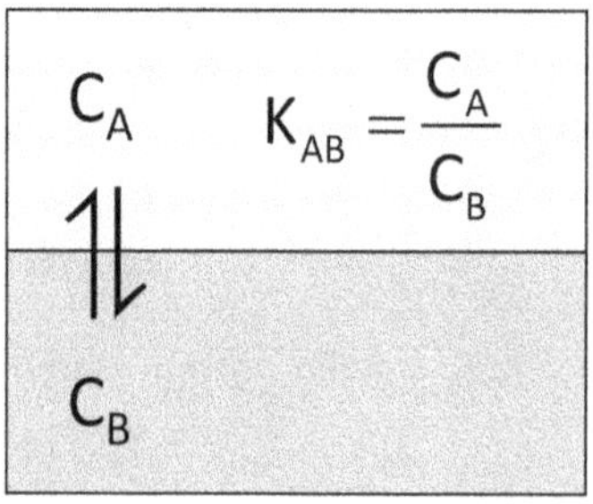

FIGURE 1.1 Illustration of the partition ratio relationship between the concentrations in two media compartments in contact at equilibrium.

Word of Warning! Partition ratios are often *unitless*, as is the case when the two concentrations are expressed in the same units, but need not necessarily be so. For example, we shall see that partition ratios for soil–water partitioning are sometimes expressed in terms of amount of chemical/mass of soil over amount of chemical/volume water, leading to non-intuitive units such as $L\,kg^{-1}$ for the $K_{Soil\text{-}W}$ partition ratio.

Worked Example 1.3

Octanol containing a chemical with a concentration of 5.00×10^{-2} *mol* L^{-1} is in contact with water at equilibrium. The chemical's concentration in the water is 7.00×10^{-5} *mol* L^{-1}. What is the chemical's octanol–water partition ratio, expressed as log K_{OW}?

K_{OW} is a *unitless* partition ratio and, as such, the units in which concentration are expressed can be any reasonable choice as long as they are the same in both the numerator and denominator. Therefore, in this example, we can directly use the concentrations expressed in *mol* L^{-1} as given, without needing to convert to *mol* m^{-3}:

$$K_{OW} = \frac{C_O}{C_W} = \frac{5.00 \times 10^{-2} mol\ L^{-1}}{7.00 \times 10^{-5} mol\ L^{-1}} = 7.14_3 \times 10^2$$

$$\log K_{OW} = \log\left(7.14_3 \times 10^2\right) = 2.854$$

Note that, when taking the logarithm of a number, the number of significant digits in the original number determines (matches) the number of significant digits *after* the decimal place in the logarithm of the number. The digits before the decimal in the logarithm reflect the order of magnitude of the number and, as pure integers, are not associated with the relative uncertainty in the quantity itself.

Word of Warning! The octanol–water partition ratio can be defined in two forms. The most common is derived experimentally from equilibrated octanol–water mixtures, in which both phases are saturated with the other solvent. This is of little concern for the aqueous phase, as the solubility of octanol is quite small in water. However, the reverse is not true, and water is appreciably soluble in octanol, such that "wet" octanol contains about 1.6 *mol* L^{-1} water. Therefore, this form of log K_{OW} should rightly be called the "wet" or "water-saturated" octanol–water partition ratio.

The other (dry) form of K_{OW} may be obtained by independent measurements of a chemical in dry octanol and pure water, and this value is useful in "thermodynamic triangle" calculations in which other partition ratios involve dry octanol, such as K_{OA}, the octanol–air partition ratio.

We may profitably rearrange the partition ratio definition to express the concentration in each compartment in terms of the equilibrated concentration in another, such as:

$$C_A = K_{AB} C_B \quad or \quad C_B = \frac{C_A}{K_{AB}}$$

Worked Example 1.4

Octanol containing a chemical with a concentration 5.00×10^{-2} *mol* L^{-1} is in equilibrium with water. The chemical has a log K_{OW} value of 2.854. What is the concentration of the chemical in the water?

Here only a simple substitution is needed, noting that $K_{OW} = 10^{\log K_{OW}}$:

$$C_W = \frac{C_O}{K_{OW}} = \frac{5.00 \times 10^{-2} mol\ L^{-1}}{10^{2.854}} = 7.00 \times 10^{-5}\ mol\ L^{-1}$$

With this very modest amount of information in hand, we can construct equilibrium partitioning models, which can be used to model the distribution of a chemical between any number of environmental compartments at equilibrium. For example, the concentration of a chemical at equilibrium between the three compartments of air, water, and soil may be compactly related through a simple mass-balance expression for the total molar amount of chemical, *m*:

$$m = m_A + m_W + m_{Soil}$$

$$m = C_A V_A + C_W V_W + C_{Soil} V_{Soil}$$

In order to reduce the problem from three variables to one variable, we can use the partition ratios to express the total content of the system in terms of one chosen concentration value. Using the air concentration as an example, we have for the same three-compartment system:

$$m = C_A V_A + \left(\frac{C_A}{K_{AW}}\right) V_W + (C_A K_{Soil-A}) V_{Soil}$$

Isolating the common factor C_A we have:

$$m = C_A \left[V_A + \left(\frac{1}{K_{AW}}\right) V_W + (K_{Soil-A}) V_{Soil} \right]$$

From here, the unknown concentration in the chosen air compartment can be isolated as:

$$C_A = \frac{m}{\left[V_A + \left(\frac{1}{K_{AW}}\right) V_W + (K_{Soil-A}) V_{Soil} \right]}$$

The concentrations in the other compartments follow from the partition ratios, and the molar amounts in each compartment from the product of compartment concentrations and volumes:

$$C_W = \frac{C_A}{K_{AW}}$$

$$C_{Soil} = C_A K_{Soil-A}$$

$$m_A = C_A V_A$$

$$m_W = C_W V_W$$

$$m_{Soil} = C_{Soil} V_{Soil}$$

Worked Example 1.5

A three-compartment system consists of air, water, and sediment, with respective volumes, V_A = 6.00×10^3 m^3, $V_W = 4.00 \times 10^2$ m^3, and $V_{Sed} = 0.400$ m^3. 1.00 *kg* of a chemical with a molar mass of 200 *g* mol^{-1} and with partition ratios of log $K_{AW} = -4.00$ and log $K_{Sed-W} = 3.00$ is released into the water. Assuming the chemical reaches equilibrium distribution between the three phases, determine the concentration and molar amounts in each phase.

We start with the mass balance expression for the system:

$$m = C_A V_A + C_W V_W + C_{Sed} V_{Sed}$$

Next, given that the two partition ratios have water as a common phase, it will be simplest to solve for the water concentration by expressing the other concentrations in terms of C_W:

$$m = (C_W K_{AW}) V_A + C_W V_W + (C_W K_{Sed-W}) V_{Sed}$$

$$m = C_W \left[K_{AW} V_A + V_W + K_{Sed-W} V_{Sed} \right]$$

$$C_W = \frac{m}{\left[K_{AW} V_A + V_W + K_{Sed-W} V_{Sed} \right]}$$

$$C_W = \frac{\left(\dfrac{1.00\,kg \times 1000 g\,kg^{-1}}{200\,g\,mol^{-1}} \right)}{\left[10^{-4.00} \times 6.00 \times 10^3 m^3 + 4.00 \times 10^2 m^3 + 10^{3.00} \times 0.400\,m^3 \right]}$$

$$C_W = \frac{5.00\,mol}{\left[6.0_0 \times 10^{-1} m^3 + 4.00 \times 10^2 m^3 + 4.0_0 \times 10^2 m^3 \right]} = 6.2_5 \times 10^{-3} mol\,m^{-3}$$

The remaining concentrations follow from the definition of the partition ratios:

$$C_A = K_{AW} C_W = 10^{-4.00} \times 6.2_5 \times 10^{-3} mol\,m^{-3} = 6.2_5 \times 10^{-7}\,mol\,m^{-3}$$

$$C_{Sed} = K_{Sed-W} C_W = 10^{3.00} \times 6.2_5 \times 10^{-3}\,mol\,m^{-3} = 6.2_5\;\; mol\,m^{-3}$$

The molar amounts follow from the product of concentration and volume of each compartment:

$$m_W = C_W V_W = 6.2_5 \times 10^{-3} mol\,m^{-3} \times 4.00 \times 10^2 m^3 = 2.5 mol$$

$$m_A = C_A V_A = 6.2_5 \times 10^{-7}\,mol\,m^{-3} \times 6.00 \times 10^3 m^3 = 3.7 \times 10^{-3} mol$$

$$m_{Sed} = C_{Sed} V_{Sed} = 6.2_5\;\; mol\,m^{-3} \times 0.400\;\; m^3 = 2.5 mol$$

Note that in this example, the actual amount of chemical in the water and the sediment is the same, even though the volume of the sediment is much smaller than that of the water compartment. This is because the sediment–water partition ratio is large, such that the concentration in the sediment is much higher. In this particular case, the volume and concentration ratios between water and soil happen to be reciprocal, so the small sediment volume is exactly balanced by a much greater sediment concentration.

Let's check this example by determining the total number of moles in the system, which should equal the amount released, 1.00 *kg*. First determine the number of moles in the system:

$$m = m_W + m_A + m_{Sed} = 2.5 mol + 3.8 \times 10^{-3} mol + 2.5 mol = 5.0\,mol$$

This corresponds to a total mass of 1 *kg*, as required:

$$Total\,mass = \frac{5.0\,mol \times 200 g\,mol^{-1}}{1000\,g\,kg^{-1}} = 1.0\,kg$$

1.2 FUGACITY-BASED CALCULATIONS

One can continue with this concentration-based approach for more complex systems. However, at this point we will benefit by introducing the concept of fugacity. Fugacity is a measure of the escaping tendency of a chemical from a medium or compartment. When it is high in one compartment with respect to another, there will be net movement of the chemical by diffusion from the higher-fugacity compartment to the lower-fugacity compartment, if diffusion is physically possible. Such diffusive transfer will continue either until equilibrium is established, some other change in the system eliminates the fugacity difference or contact between the two compartments is interrupted.

Fugacity is related to concentration of a given chemical by a simple constant of proportionality, the fugacity capacity Z, with unit mol Pa^{-1} m^{-3}:

$$\boxed{C = Z\,f}$$

The greater the fugacity capacity of a given medium, the less fugacity increases with concentration in that medium. Thus, a medium with a high fugacity capacity tends to be a "sink" for a chemical, since it can uptake a relatively large amount of chemical (i.e., can achieve a high concentration) without generating a correspondingly relatively high driving force for transfer to other media (i.e., a high fugacity).

Word of Warning! The fugacity capacity, or Z-value, is a property of both a particular chemical and of a given environment. For a given medium, such as water, the fugacity capacity is different and unique for every chemical. Also, for a given chemical, such as phenol, its fugacity capacity in every medium is different.

Worked Example 1.6

What is the fugacity capacity of a chemical in water if its concentration in water is 2.00 x 10^{-2} *mol* m^{-3} and its fugacity in water is 50.0 *Pa*?

Rearrangement of the concentration–fugacity relation yields:

$$Z_W = \frac{C_W}{f_W}$$

$$Z_W = \frac{2.00 \times 10^{-2} mol\ m^{-3}}{50.0\,Pa} = 4.00 \times 10^{-4}\ mol\,Pa^{-1} m^{-3}$$

A key attribute of fugacity is that it can be used as an equilibrium criterion. This is because all environmental compartments that are at equilibrium have the same fugacity, a condition called "equifugacity". Thus, although the concentrations of two compartments at equilibrium may be significantly different, their fugacity will *always* be the same. Figure 1.2 illustrates this relationship for two compartments at equilibrium.

When fugacity is viewed as an escaping tendency, two equifugacity compartments in contact are such that the rate at which the chemical escapes from compartment *A* into compartment *B* is exactly equal to the rate of escape of the chemical from compartment *B* into compartment *A*. This is a direct reflection of the fact that the chemical equilibrium is a dynamic state, with continuous and reciprocal transfer of a chemical between the equilibrating compartments.

Thus, for two compartments *A* and *B* in contact at equilibrium, we have a powerful means by which to express the equilibrium condition:

$$f_A^{Eq} = f_B^{Eq}$$

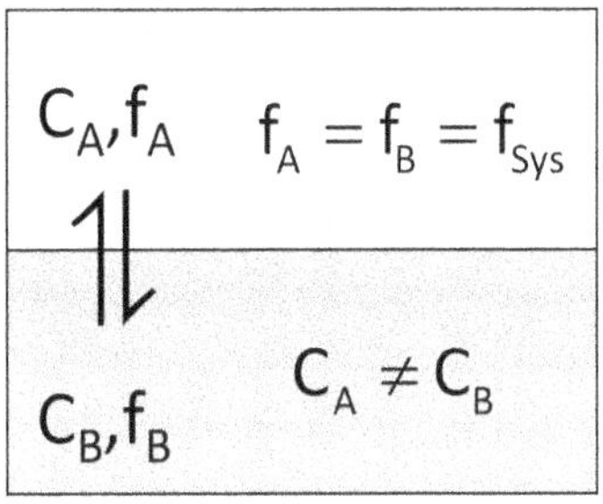

FIGURE 1.2 Illustration of the equifugacity condition for two media compartments at equilibrium.

We can extend this type of relationship to as many compartments at equilibrium as are appropriate to the system in question. For n compartments at equilibrium, each may have a unique concentration, but all will have the same fugacity:

$$f_1^{Eq} = f_2^{Eq} = f_3^{Eq} = f_4^{Eq} = f_5^{Eq} = \ldots = f_n^{Eq} = f_{Sys}$$

Here, we denote the prevailing fugacity of the equilibrium system as f_{Sys}, which applies to all compartments at equilibrium.

Introducing fugacity also allows for a very powerful restatement of the equilibrium partition ratio:

$$\boxed{K_{AB} = \frac{C_A}{C_B} = \frac{Z_A f_{Sys}}{Z_B f_{Sys}} = \frac{Z_A}{Z_B}}$$

That is, the partition ratio may be recognized as the ratio of the fugacity capacities of two compartments. In this way, a fugacity capacity is like "half a partition ratio". It is by virtue of introducing the equilibrium criterion of equifugacity that we gain this ability to "dissect" partition ratios and deal with individual environmental compartments as separate but interacting entities. This ability proves to be much valuable when modelling very complicated systems and greatly reduces the number of equations needed to solve the problem.

Multimedia calculations and models that involve closed systems in which all environmental compartments are at equilibrium (equifugacity) are termed "Level I" systems in the "Mackay" fugacity-based modelling nomenclature (see Figure 1.3). In such model systems, a chemical is "allowed" to distribute itself among the various compartments and it will concentrate in those composed of media with the highest fugacity capacities.

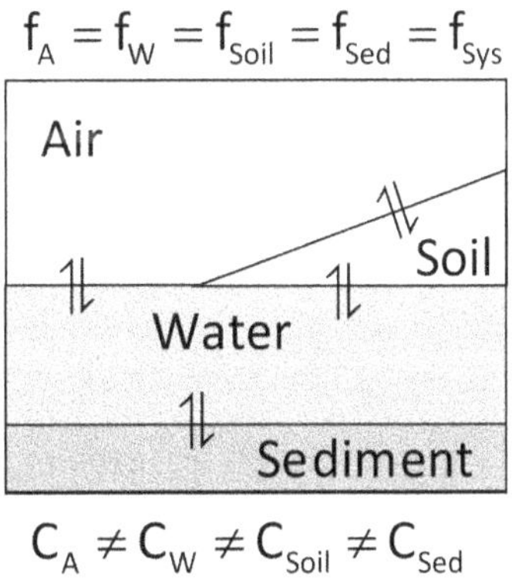

FIGURE 1.3 Four-compartment Level I closed system at equilibrium and therefore at equifugacity.

Worked Example 1.7

The fugacity capacities for a chemical in octanol and water are $Z_O = 3.20 \times 10^{-1}$ *mol Pa*$^{-1}$ *m*$^{-3}$ and $Z_W = 4.00 \times 10^{-4}$ *mol Pa*$^{-1}$ *m*$^{-3}$. What is the octanol–water partition ratio for this chemical at equilibrium?

Here we "reconstruct" the partition ratio from the fugacity capacities:

$$K_{OW} = \frac{Z_O}{Z_W}$$

$$K_{OW} = \frac{3.20 \times 10^{-1} mol\ Pa^{-1} m^{-3}}{4.00 \times 10^{-4} mol\ Pa^{-1} m^{-3}} = 800._0$$

$$\log K_{OW} = 2.903$$

Note that, as expected, the concentration in the octanol phase is 800 times that of water, reflecting its much higher fugacity capacity with respect to this chemical.

The expression of mass balance for air, water, and soil at equilibrium in terms of overall system fugacity f_{Sys} is relatively simple, and immediately reduces the equation from three variables (in this case) to one:

$$m = m_A + m_W + m_{Soil}$$

$$m = C_A V_A + C_W V_W + C_{Soil} V_{Soil}$$

$$m = Z_A f_{Sys} V_A + Z_W f_{Sys} V_W + Z_{Soil} f_{Sys} V_{Soil}$$

$$m = \left(Z_A V_A + Z_W V_W + Z_{Soil} V_{Soil}\right) f_{Sys}$$

Rearrangement of this expression leads to an expression for determining the fugacity of our system of three compartments at equilibrium:

$$f_{Sys} = \frac{m}{Z_A V_A + Z_W V_W + Z_{Soil} V_{Soil}}$$

More generally, for n compartments at equilibrium, we have:

$$\boxed{f_{Sys} = \frac{m}{\sum_{i=1}^{n} Z_i V_i}}$$

Once the system fugacity is determined, concentrations in all compartments at equilibrium are easily determined by the product of the fugacity capacity and the prevailing fugacity:

$$\boxed{C_i = Z_i f_{Sys}}$$

1.3 ESTIMATING FUGACITY CAPACITIES IN AIR, WATER, AND OCTANOL

Developing a Level I model starts with the determination of the fugacity capacities (Z-values) for the chemical(s) in question in the various media compartments. The simplest of these to estimate is the fugacity capacity of a chemical in air. Fortunately, in most environmentally relevant cases, gas-phase fugacity is essentially equal to the vapour pressure of a chemical at equilibrium with its pure liquid phase. (Exceptions occur at high temperatures and pressures, and with dissociating chemicals.) Therefore, we can establish the fugacity capacity of any non-interacting chemical in air rather simply, assuming ideal gas behaviour, via:

$$Z_A = \frac{C_A}{f_{Sys}} = \frac{\left(\frac{m_A}{V_A}\right)}{P_A} = \frac{\left(\frac{P_A}{RT}\right)}{P_A} = \frac{1}{RT}$$

$$\boxed{Z_A = \frac{1}{RT}}$$

Worked Example 1.8

What is the fugacity capacity of a non-interacting chemical in air at 27.5°C?

Straightforward substitution with conversion from degrees Celsius to Kelvin yields:

$$Z_A = \frac{1}{RT} = \frac{1}{8.31446\,m^3 Pa\,K^{-1} mol^{-1} \times (27.5 + 273.15)\,K} = 4.000 \times 10^{-4}\,mol\,Pa^{-1}\,m^{-3}$$

The fugacity capacity of air is temperature dependent and should be calculated for the appropriate temperature of the system. Figure 1.4 shows the variation in the fugacity capacity of any (ideal gas) chemical in air as a function of temperature over the environmentally relevant range of −40 to +40°C:

Although Z_A varies by about 30% over this wide temperature range, much modelling work is done assuming "room" temperature, at which Z_A is very close to 4.0×10^{-4} $mol\ Pa^{-1}\ m^{-3}$, and this value is often used when temperature is not specified.

The fugacity capacity of a chemical in all environmentally relevant aqueous phases is usually expressed in terms of the fugacity capacity in water, Z_W. This fugacity capacity is most conveniently

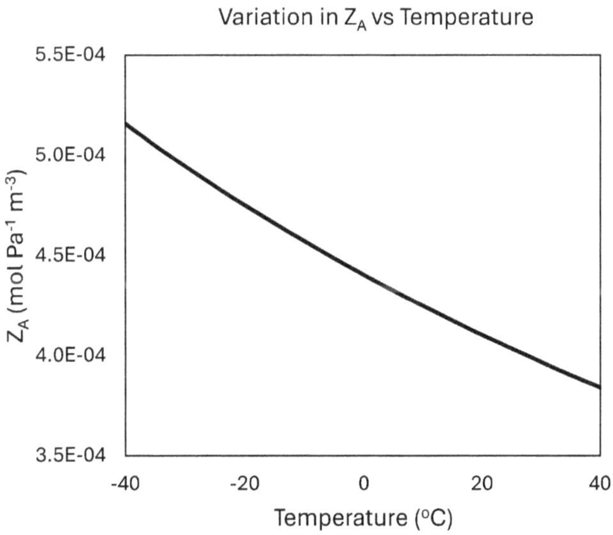

FIGURE 1.4 Plot of the variation in the fugacity capacity of air as a function of temperature in the range −40 to 40°C.

obtained from the air–water partition ratio K_{AW}. Taking advantage of the "half partition ratio" nature of fugacity capacities, we can immediately write:

$$K_{AW} = \frac{Z_A}{Z_W}$$

or

$$\boxed{Z_W = \frac{Z_A}{K_{AW}}}$$

Worked Example 1.9

A chemical has a value of log $K_{AW} = -4.22$. What is its fugacity capacity in water at room temperature?

The fugacity capacity in air for all chemicals is the same under the ideal gas assumption. We assume room temperature and therefore $Z_A = 4.0 \times 10^{-4}\ mol\ Pa^{-1}\ m^{-3}$. With this in hand, we can directly calculate Z_W:

$$Z_W = \frac{Z_A}{K_{AW}} = \frac{4.0\times10^{-4} mol\,Pa^{-1}m^{-3}}{10^{-4.22}} = \frac{4.0\times10^{-4} mol\,Pa^{-1}m^{-3}}{6.0_3\times10^{-5}} = 6.6\,mol\,Pa^{-1}m^{-3}$$

The Henry's Law constant is closely related to the air–water partition ratio. It is the constant of proportionality between a chemical's partial vapour pressure above water and its corresponding concentration in water at low enough chemical concentrations that "self" interactions between two solute molecules can be ignored.

$$H = \frac{P_A}{C_W}$$

The Henry's Law constant may also be calculated from the ratio of the saturation vapour pressure of the chemical in air over the saturation concentration of the chemical in water:

$$H = \frac{P_A^{Sat}}{C_W^{Sat}}$$

The following is the relationship between K_{AW} and H, where we assume ideal gas behaviour:

$$K_{AW} = \frac{C_A}{C_W} = \frac{\left(\frac{n_A}{V_A}\right)}{C_W} = \frac{\left(\frac{P_A}{RT}\right)}{C_W} = \left(\frac{P_A}{C_W}\right)\times\frac{1}{RT} = (H)\times\frac{1}{RT} = \frac{H}{RT}$$

$$\boxed{K_{AW} = \frac{H}{RT}}$$

Using this equation relating K_{AW} and H, and the equation for Z_A introduced above, we can also determine the fugacity capacity of a chemical in water in terms of the chemical's Henry's Law constant:

$$Z_W = \frac{Z_A}{K_{AW}} = \frac{\left(\frac{1}{RT}\right)}{\left(\frac{H}{RT}\right)} = \frac{1}{H}$$

$$\boxed{Z_W = \frac{1}{H}}$$

We see that the fugacity capacity of water is simply the inverse of the Henry's Law constant.

Word of Warning! Take care as various units are used for air–water partition ratios and Henry's Law constants, and the relationships themselves may be expressed in inverse form as well. For example, pressure is commonly reported in both Pascals and atmospheres, and Henry's Law constants may be found as P_A/C_W or C_W/P_A. Use the units as your guide. You can only take the direct inverse of a Henry's Law constant in $Pa\ m^3\ mol^{-1}$ units to arrive at a fugacity capacity of water in the units of $mol\ Pa^{-1}\ m^{-3}$, which are the units used throughout this book. If the Henry's Law constant for a chemical of interest is in other units, you will need to include one or more unit conversions in your calculation and may even need to invert it as well!

Worked Example 1.10

The same chemical as in Worked Example 1.9 has a Henry's Law constant of $H = 1.50 \times 10^{-6}\ atm\ m^3\ mol^{-1}$. Use this value to recalculate its fugacity capacity in water.

Since we have the Henry's Law constant, we can directly calculate Z_W. However, note that H is not provided in units of $Pa\ m^3\ mol^{-1}$, so attention must be given to unit conversion from atmospheres to pascals:

$$Z_W = \frac{1}{H}$$

$$Z_W = \frac{1}{1.50 \times 10^{-6}\, atm\, m^3\, mol^{-1} \times 101325\, Pa\, atm^{-1}} = 6.58\, mol\, Pa^{-1}\, m^{-3}$$

In the event that the Henry's Law constant is not immediately available, the fugacity capacity of water can also be determined from the water solubility of the chemical and its saturation vapour pressure, the inverse of one way of expressing H, just in disguise:

$$\boxed{Z_W = \frac{C_W^{Sat}}{P_A^{Sat}}}$$

Worked Example 1.11

The water solubility of benzene (molar mass 78.11 $g\ mol^{-1}$) is 1.780 $g\ L^{-1}$ and its saturation vapour pressure is $1.27 \times 10^4\ Pa$. What is the fugacity capacity of benzene in water?

The problem is a direct application of the expression for Z_W given above but requires some unit conversion from grams to moles and from litres to cubic metres to obtain the fugacity capacity in units of $mol\ Pa^{-1}\ m^{-3}$.

$$Z_W = \frac{C_W^{Sat}}{P_A^{Sat}}$$

$$Z_W = \frac{\left(\dfrac{1.780\, g\, L^{-1}}{78.11 g\, mol^{-1}}\right) \times 10^3 L\, m^{-3}}{1.27 \times 10^4\, Pa}$$

$$Z_W = 1.79 \times 10^{-3} mol\, Pa^{-1}\, m^{-3}$$

When one uses this approach involving the pure-phase vapour pressure of a substance, one must be mindful of the fact that some chemicals are solid at the temperature of interest, in which case the subcooled liquid saturation vapour pressure $P_A^{Sat-SCL}$ must be calculated and used instead:

$$\boxed{Z_W\,(solids) = \frac{C_W^{Sat}}{P_A^{Sat-SCL}}}$$

The ratio of the solid-state saturation vapour pressure over the subcooled liquid saturation vapour pressure is termed the fugacity ratio F:

$$\boxed{F = \frac{P_A^{Sat}}{P_A^{Sat-SCL}}}$$

Note that F is always less than or equal to 1, such that the subcooled liquid saturation vapour pressure will always be greater than or equal to the solid-state value. This makes sense, since a theoretical subcooled liquid molecule does not have to break solid-state intermolecular interactions to move from the condensed state into the gaseous state, and therefore has more kinetic energy available, which manifests as an increased vapour pressure.

Word of Warning! Experimentally measured vapour pressures are usually those of the solid if the measurement is carried out below the melting point of the compound, whereas QSAR estimates may give a subcooled liquid vapour pressure estimate. One should always compare the melting point of the chemical to the temperature of interest and if the latter is the lower, check whether you have the solid or subcooled liquid vapour pressure on hand, making the conversion if necessary.

Worked Example 1.12

The saturation vapour pressure of solid naphthalene (molar mass 128.18 $g\,mol^{-1}$) is 5.38 Pa and its fugacity ratio is $F = 0.2825$. Determine the subcooled liquid saturation vapour pressure of naphthalene and use this value with its water solubility of 31.0 $g\,m^{-3}$ to determine Z_W for naphthalene.

First determine the subcooled liquid saturation vapour pressure from the fugacity ratio and the solid-state saturation vapour pressure:

$$F = \frac{P_A^{Sat}}{P_A^{Sat-SCL}}$$

$$P_A^{Sat-SCL} = \frac{P_A^{Sat}}{F} = \frac{5.38\ Pa}{0.2825} = 19.0_4 Pa$$

Now use this value with the water solubility of naphthalene to determine the fugacity capacity in water:

$$Z_W = \frac{C_W^{Sat}}{P_A^{Sat-SCL}}$$

$$Z_W = \frac{\left(\dfrac{31.0\,g\,m^{-3}}{128.18\,g\,mol^{-1}}\right)}{19.0_4\ Pa} = 1.27\times10^{-2} mol\,Pa^{-1}m^{-3}$$

The fugacity ratio is often estimated at a given temperature $T(K)$ from the chemical's melting point in Kelvin by the following expression which assumes Walden's Rule:

$$\boxed{\ln F \simeq 6.79\left(1-\frac{T_M}{T(K)}\right)}$$

At 298 K and converting to base-10 logarithms, the following simplification is possible:

$$\boxed{\log F \simeq 0.01\left(298 - T_M(K)\right)}$$

Worked Example 1.13

Use the Walden's Rule and the base-10 simplification expressions given above to calculate the fugacity ratio for naphthalene at 25.0°C, given a melting point of 80.5°C.

Using the natural logarithm expression first, and taking care to convert to Kelvin, we have:

$$\ln F \simeq 6.79\left(1-\frac{T_M}{T(K)}\right)$$

$$\ln F \simeq 6.79\left(1-\frac{(80.5+273.15)K}{(25.0+273.15)K}\right) = -1.26_4$$

$$F \simeq \exp\left(-1.26_4\right) = 0.283$$

Now employing the base 10 logarithm expression:

$$\log F \simeq 0.01\left(298 - T_M(K)\right)$$

$$\log F \simeq 0.01\left(298-(80.5+273.15)\right) = -0.55_7$$

$$F \simeq 10^{-0.55_7} = 0.28$$

Octanol (n-octanol) is used as a surrogate for all of, or a fraction of, many non-aqueous (non-polar) environmental media. The fugacity capacity of a chemical in octanol is most conveniently obtained via the octanol–water partition ratio, K_{OW}, which is usually readily available or can be estimated with acceptable accuracy. Again, by taking advantage of the "half partition ratio" nature of fugacity capacities, we can derive one fugacity capacity from another via the partition ratio:

$$K_{OW} = \frac{Z_O}{Z_W}$$

$$\boxed{Z_O = K_{OW} Z_W}$$

Worked Example 1.14

A chemical has a log K_{OW} value of 3.25. What is its octanol fugacity capacity if Z_W = 6.58 mol Pa^{-1} m^{-3}?

The calculation is straightforward:

$$Z_O = K_{OW} Z_W = 10^{3.25} \times 6.58\ mol\, Pa^{-1} m^{-3} = 1.2 \times 10^4\ mol\, Pa^{-1} m^{-3}$$

With these three fugacity capacities in hand, we can solve for the system fugacity and then the concentrations and molar amounts in each of the three compartments of air, water, and octanol, according to the relations introduced above:

$$f_{Sys} = \frac{m}{\sum_{i=1}^{n} Z_i V_i} = \frac{m}{Z_A V_A + Z_W V_W + Z_O V_O}$$

$$C_i = Z_i f_{Sys}$$

$$m_i = C_i V_i$$

Worked Example 1.15

1.00 *g* of a chemical of molar mass 200 $g\ mol^{-1}$ is introduced into a closed vessel containing 250 *mL* of octanol and 250 *mL* of water. There is a 100 *mL* volume of air above the liquid. The fugacity capacities of the chemical in air, water, and octanol are $Z_A = 4.00 \times 10^{-4}\ mol\ Pa^{-1}\ m^{-3}$, $Z_W = 6.58\ mol\ Pa^{-1}\ m^{-3}$ and $Z_O = 1.17 \times 10^{4}\ mol\ Pa^{-1}\ m^{-3}$. What are the concentrations and mole percentages of the chemical in each medium at equilibrium?

We have the fugacity capacities and compartment volumes, so we can calculate the system fugacity directly. However, the volumes are provided in *mL* and it will "clean up" the equations a bit if we first convert the compartment volumes to m^3 before calculating the system fugacity:

$$V_A = \frac{100.\,mL}{10^6\,mL\,m^{-3}} = 1.00 \times 10^{-4}\,m^3$$

$$V_W = \frac{250.\,mL}{10^6\,mL\,m^{-3}} = 2.50 \times 10^{-4}\,m^3$$

$$V_O = \frac{250.\,mL}{10^6\,mL\,m^{-3}} = 2.50 \times 10^{-4}\,m^3$$

Now, using these compartment volumes and the fugacity capacities, the system fugacity is:

$$f_{Sys} = \frac{m}{Z_A V_A + Z_W V_W + Z_O V_O}$$

The numerator is:

$$m = \frac{1.00\,g}{200.\,g\,mol^{-1}} = 5.00_0 \times 10^{-3}\,mol$$

The denominator is:

$$\begin{aligned}
&Z_A V_A + Z_W V_W + Z_O V_O \\
&= 4.00 \times 10^{-4}\,mol\,Pa^{-1}\,m^{-3} \times 1.00 \times 10^{-4}\,m^3 + 6.58\,mol\,Pa^{-1}\,m^{-3} \times 2.50 \times 10^{-4}\,m^3 \\
&\quad + 1.17 \times 10^4\,mol\,Pa^{-1}\,m^{-3} \times 2.50 \times 10^{-4}\,m^3 \\
&= 4.00 \times 10^{-8}\,mol\,Pa^{-1} + 1.64_5 \times 10^{-3}\,mol\,Pa^{-1} + 2.92_5\,mol\,Pa^{-1} \\
&= 2.92_7\,mol\,Pa^{-1}
\end{aligned}$$

Finally, the fugacity is:

$$f_{Sys} = \frac{m}{Z_A V_A + Z_W V_W + Z_O V_O} = \frac{5.00_0 \times 10^{-3}\,mol}{2.92_7\,mol\,Pa^{-1}} = 1.70_8 \times 10^{-3}\,Pa$$

The concentrations in each compartment are:

$$C_A = Z_A f_{Sys} = 4.00 \times 10^{-4} mol\,Pa^{-1}\,m^{-3} \times 1.70_8 \times 10^{-3}\,Pa = 6.83_4 \times 10^{-7}\,mol\,m^{-3}$$

$$C_W = Z_W f_{Sys} = 6.58\,mol\,Pa^{-1}\,m^{-3} \times 1.70_8 \times 10^{-3}\,Pa = 1.12_4 \times 10^{-2}\,mol\,m^{-3}$$

$$C_O = Z_O f_{Sys} = 1.17 \times 10^{4}\,mol\,Pa^{-1}\,m^{-3} \times 1.70_8 \times 10^{-3}\,Pa = 19.9_9\,mol\,m^{-3}$$

The mole percentages of chemical in each compartment are given by:

$$\%x_A = \frac{C_A V_A}{m} \times 100\% = \left(\frac{6.83_4 \times 10^{-7}\,mol\,m^{-3} \times 1.00 \times 10^{-4}\,m^3}{5.00_0 \times 10^{-3}\,mol} \right) \times 100\% = 1.36_7 \times 10^{-6}\,\%$$

$$\%x_W = \frac{C_W V_W}{m} \times 100\% = \left(\frac{1.12_4 \times 10^{-2}\,mol\,m^{-3} \times 2.50 \times 10^{-4}\,m^3}{5.00_0 \times 10^{-3}\,mol} \right) \times 100\% = 5.62_1 \times 10^{-2}\,\%$$

$$\%x_O = \frac{C_O V_O}{m} \times 100\% = \left(\frac{19.9_9\,mol\,m^{-3} \times 2.50 \times 10^{-4}\,m^3}{5.00_0 \times 10^{-3}\,mol} \right) \times 100\% = 99.9_4\%$$

Note that the relatively high fugacity capacity of the chemical in octanol results in virtually all of the chemical partitioning to the octanol phase.

1.4 PARTITIONING TO ORGANIC CARBON

Many environmentally relevant phases such as soils and sediments are composed of a combination of organic and inorganic materials. From a partitioning perspective, the organic portion is usually dominant, while the inorganic phase is a minor contributor and, in some cases, is relegated to the category of inert "filler" material. Partitioning from water to the organic portion of such materials can be expressed in terms of the organic carbon–water partition ratio, $K_{OC\text{-}W}$, commonly known as K_{OC}. Since the latter name can be somewhat misleading to students, we will use the more explicit $K_{OC\text{-}W}$ in this book for clarity.

Word of Warning! Students should bear in mind that the use of the symbol $K_{OC\text{-}W}$ is not the norm in professional usage, where only K_{OC} will be encountered.

$K_{OC\text{-}W}$ is normally expressed in units of $L\,kg^{-1}$ in keeping with its definition as:

$$K_{OC\text{-}W}\left(L\ kg^{-1}\right) = \left(\frac{amount\ of\ chemical/kg\,OC}{amount\ of\ chemical/L\ water} \right)$$

Here, the kg in the units of $K_{OC\text{-}W}$ pertains to mass of organic carbon (OC) in the sample, and the litre pertains to the volume of water. The amount of chemical may be expressed in mass or mole units, as long as the same unit choice is made for the amount in both phases.

In order to express $K_{OC\text{-}W}$ in *unitless* form, it is necessary to convert the mass of OC to the corresponding volume of OC using the OC density ρ_{OC}. Thus, we have the *unitless* $K_{OC\text{-}W}$:

$$K_{OC\text{-}W}(unitless) = K_{OC\text{-}W}\left(L\ kg^{-1}\right) \times \rho_{OC}\left(kg\,L^{-1}\right)$$

A factor of 1000 $L\,m^{-3}$ must be introduced to allow for expression of the density of OC in SI units of $kg\,m^{-3}$:

$$K_{OC-W}(unitless) = K_{OC-W}\left(L\ kg^{-1}\right) \times \left(\frac{\rho_{OC}\left(kg\,m^{-3}\right)}{1000\,L\,m^{-3}}\right)$$

Generally, it is assumed that there is a straightforward, albeit approximate, relation between K_{OC-W} and *unitless* K_{OW}, which often takes the following form:

$$K_{OC-W}\left(L\ kg^{-1}\right) = \left(0.41\,L\ kg^{-1}\right)K_{OW}$$

or

$$\boxed{K_{OC-W}(unitless) = \left(0.41\,L\ kg^{-1}\right)K_{OW} \times \left(\frac{\rho_{OC}\left(kg\,m^{-3}\right)}{1000\,L\,m^{-3}}\right)}$$

This equation may be called the Karickhoff relation for reference, after Karickhoff who published a detailed study that led to this equation (Karickhoff, 1981). Estimates of K_{OC-W} from this approach were found to be accurate to within a factor of two, with an absolute average deviation in log K_{OC-W} of 0.21 for non-polar and 0.31 for polar molecules.

Assuming a density of OC of 1000 $kg\,m^{-3}$, and the more conservative error associated with polar molecules, the corresponding relationship between log (K_{OC-W} (*unitless*)) and log K_{OW} is:

$$\boxed{\log\left(K_{OC-W}(unitless)\right) = \log K_{OW} - (0.39 \pm 0.31)}$$

Seth et al. re-examined this question, included many more chemicals, and suggested that a conversion factor of 0.35 $L\,kg^{-1}$ is more appropriate, also with a fairly wide margin of uncertainty. This work also provided positive and negative error estimates (Seth et al., 1999). The "Seth" relationship between K_{OC-W} and K_{OW} is:

$$\boxed{K_{OC-W}(unitless) = \left(0.35^{+0.54}_{-0.21}\,L\ kg^{-1}\right)K_{OW} \times \left(\frac{\rho_{OC}\left(kg\,m^{-3}\right)}{1000\,L\,m^{-3}}\right)}$$

If we assume a density of organic carbon of 1000 $kg\,m^{-3}$, we can generate this relationship in log form as:

$$\boxed{\log\left(K_{OC-W}(unitless)\right) = \log K_{OW} - (0.46 \pm 0.39)}$$

It is clear that the inclusion of K_{OC-W} derived from K_{OW} in any modelling scenario constitutes a significant source of uncertainty. We shall see below that many environmental phases are modelled using this partition ratio or a fugacity capacity derived from it and are therefore subject to at least this degree of uncertainty. In particular, K_{OC-W} is very important to agro-chemists dealing with pesticides, for which experimental determinations are made for the site of interest. As a result, the nature of organic carbon in soils has been studied extensively.

We shall use the Seth expression in this workbook unless otherwise noted, although as demonstrated in the Worked Example 1.16, the difference between the estimates from the two approaches is much smaller than the uncertainty associated with the values themselves.

Word of Warning! The input for most of the models available from the Canadian Environmental Modelling Centre (CEMC) website (Trent University), to which this book often refers, requires you to enter a value for K_{OC}, which amounts to choosing either the Karickhoff or Seth proportionality constant. Always check to see which is being used if you are going to compare your own calculation results against a model such as these, as the choice of either 0.41 or 0.35 will influence the outcome sufficiently to cause you to question your calculation. Note also that most of the older VisualBasic versions of the CEMC models assume the Karickhoff relationship value of 0.41, but do not indicate this explicitly. One can determine which value is being used by calculating $0.35 \times 10^{\log K_{ow}}$ and $0.41 \times 10^{\log K_{ow}}$ and comparing the result to the value of K_{OC} given in the model's results.

Worked Example 1.16

Use the Karickhoff and Seth relationships to calculate the *unitless* log $K_{OC\text{-}W}$ for a chemical with log K_{OW} = 2.60, assuming an organic carbon of density of 1000 $kg\ m^{-3}$.

First evaluate K_{OW}:

$$K_{OW} = 10^{2.60} = 3.9_8 \times 10^2$$

Now evaluate the *unitless* $K_{OC\text{-}W}$ using the Karickhoff relation and the *OC* density:

$$K_{OC\text{-}W}(unitless) = 0.41L\ kg^{-1} \times K_{OW} \times \left(\frac{\rho_{OC}\left(kg\,m^{-3}\right)}{1000\,L\,m^{-3}} \right)$$

$$K_{OC\text{-}W}(unitless) = 0.41L\ kg^{-1} \times 3.9_8 \times 10^2 \times \left(\frac{1000\,kg\,m^{-3}}{1000\,L\,m^{-3}} \right)$$

$$K_{OC\text{-}W}(unitless) = 1.6_3 \times 10^2$$

$$\log\left(K_{OC\text{-}W}(unitless)\right) = \log\left(1.6_3 \times 10^2\right) = 2.21(\pm 0.31)$$

The *unitless* $K_{OC\text{-}W}$ using the Seth relationship is:

$$K_{OC\text{-}W}(unitless) = 0.35L\ kg^{-1} \times K_{OW} \times \left(\frac{\rho_{OC}\left(kg\,m^{-3}\right)}{1000\,L\,m^{-3}} \right)$$

$$K_{OC\text{-}W}(unitless) = 0.35L\ kg^{-1} \times 3.9_8 \times 10^2 \times \left(\frac{1000\,kg\,m^{-3}}{1000\,L\,m^{-3}} \right)$$

$$K_{OC\text{-}W}(unitless) = 1.3_9 \times 10^2$$

$$\log\left(K_{OC\text{-}W}(unitless)\right) = \log\left(1.3_9 \times 10^2\right) = 2.14(\pm 0.39)$$

Alternatively, since the density of organic carbon is given as 1000 $kg\ m^{-3}$, we could have directly used the "condensed" log-log relationships given above, both of which assume this same density for organic carbon.

For the Karickhoff relation:

$$\log\left(K_{OC\text{-}W}(unitless)\right) = \log K_{OW} - 0.39(\pm 0.31)$$

$$\log\left(K_{OC\text{-}W}(unitless)\right) = 2.60 - 0.39(\pm 0.31)$$

$$\log\left(K_{OC\text{-}W}(unitless)\right) = 2.21(\pm 0.31)$$

For the Seth relation:

$$\log\left(K_{OC\text{-}W}(unitless)\right) = \log K_{OW} - 0.46(\pm 0.39)$$

$$\log\left(K_{OC\text{-}W}(unitless)\right) = 2.60 - 0.46(\pm 0.39)$$

$$\log\left(K_{OC\text{-}W}(unitless)\right) = 2.14(\pm 0.39)$$

Notice that log $K_{OC\text{-}W}$ will always be somewhat less than log K_{OW}, reflecting the fact that organic carbon has a smaller fugacity capacity than octanol for all chemicals by a factor of about one third (0.41 or 0.35 here). As well, it is clear that both relationships give similar results and that the associated errors are significantly greater than the difference between the Karickhoff and Seth estimates.

As noted above, most phases that contain organic carbon are not "pure" organic carbon, but rather are composed of a certain fraction of organic carbon as well as inorganic material. For example, a soil might be 2% organic carbon content by volume, with the remainder considered to be inert inorganic material to which the chemical will not partition to any significant extent (ignoring water and air for simplicity). Correspondingly, the volume fraction of organic carbon is $v_{Soil}^{f-OC} = 0.02$. The soil–water partition ratio for such a soil comprises only organic carbon and inert material is then K_{OC-W} scaled by the volume fraction of organic carbon content in the soil:

$$K_{Soil-W} = v_{Soil}^{f-OC} K_{OC-W}$$

Substituting the expression for the *unitless* K_{OC-W} in terms of the Seth relation developed above, we arrive at an expression for the *unitless* K_{Soil-W} in terms of the easily obtained or estimated K_{OW}, and the volume fraction and density of organic carbon:

$$\boxed{K_{Soil-W}(unitless) = v_{Soil}^{f-OC} \times 0.35\,L\ kg^{-1} \times K_{OW} \times \left(\frac{\rho_{OC}\left(kg\,m^{-3}\right)}{1000\,L\,m^{-3}}\right)}$$

Since the water phase "half" of K_{Soil-W} is considered pure, we can easily establish a definition for the fugacity capacity of this soil by multiplying both sides by Z_W to obtain:

$$\boxed{Z_{Soil} = K_{Soil-W}(unitless) \times Z_W = v_{Soil}^{f-OC} \times 0.35\,L\ kg^{-1} \times K_{OW} \times Z_W \times \left(\frac{\rho_{OC}\left(kg\,m^{-3}\right)}{1000\,L\,m^{-3}}\right)}$$

Thus, we arrive at a means to establish the fugacity capacity of soil in terms of the relatively easily obtained K_{OW} and Z_W.

In the event that the organic carbon content of the soil is measured in terms of a mass fraction (i.e., mass of *OC* per mass of soil), a further conversion factor involving the density of *OC* and soil is needed. The volume and mass fractions are related in the following way:

$$v_{Soil}^{f-OC} = \frac{V_{OC}}{V_{Soil}} = \frac{mass_{OC}(kg)/\rho_{OC}(kg\,m^{-3})}{mass_{Soil}(kg)/\rho_{Soil}(kg\,m^{-3})} = m_{Soil}^{f-OC}\left(\frac{\rho_{Soil}(kg\,m^{-3})}{\rho_{OC}(kg\,m^{-3})}\right)$$

Use of the mass fraction as a measure of *OC* in soil imparts an appealing mathematical advantage, in that substitution of mass fraction for volume fraction allows one to cancel the density of *OC*, leaving only the more easily estimated density of soil. Again, using the Seth relation as an example in soils:

$$K_{Soil-W}(unitless) = v_{Soil}^{f-OC} \times 0.35\,L\ kg^{-1} \times K_{OW} \times \left(\frac{\rho_{OC}\left(kg\,m^{-3}\right)}{1000\,L\,m^{-3}} \right)$$

$$K_{Soil-W}(unitless) = m_{Soil}^{f-OC} \left(\frac{\rho_{Soil}(kg\,m^{-3})}{\rho_{OC}(kg\,m^{-3})} \right) \times 0.35\,L\ kg^{-1} \times K_{OW} \times \left(\frac{\rho_{OC}\left(kg\,m^{-3}\right)}{1000\,L\,m^{-3}} \right)$$

$$\boxed{K_{Soil-W}(unitless) = m_{Soil}^{f-OC} \times 0.35\,L\ kg^{-1} \times K_{OW} \times \left(\frac{\rho_{Soil}(kg\,m^{-3})}{1000\,L\,m^{-3}} \right)}$$

The corresponding fugacity capacity expression is:

$$\boxed{Z_{Soil} = m_{Soil}^{f-OC} \times 0.35\,L\ kg^{-1} \times K_{OW} \times Z_W \times \left(\frac{\rho_{Soil}\left(kg\,m^{-3}\right)}{1000\,L\,m^{-3}} \right)}$$

In general, for any medium of density ρ_{Medium} for which the partitioning is deemed to be only to its organic carbon content, and for which the organic carbon fractions are given either in terms of volume fraction v_{Medium}^{f-OC} or mass fraction m_{Medium}^{f-OC}, we have:

$$\boxed{Z_{Medium} = v_{Medium}^{f-OC} \times 0.35\,L\ kg^{-1} \times K_{OW} \times Z_W \times \left(\frac{\rho_{OC}\left(kg\,m^{-3}\right)}{1000\,L\,m^{-3}} \right)}$$

or

$$\boxed{Z_{Medium} = m_{Medium}^{f-OC} \times 0.35\,L\ kg^{-1} \times K_{OW} \times Z_W \times \left(\frac{\rho_{Medium}\left(kg\,m^{-3}\right)}{1000\,L\,m^{-3}} \right)}$$

Worked Example 1.17

Estimate K_{Soil-W} and Z_{Soil} for a chemical with a Henry's Law constant $H = 2.00 \times 10^{-3}\ atm\ m^3\ mol^{-1}$ and $\log K_{OW} = 4.30$. Assume an organic carbon density of 1000 $kg\ m^{-3}$ and an organic carbon volume fraction $v_{OC}^{f} = 0.025$.

K_{Soil-W} can be determined immediately:

$$K_{Soil-W}(unitless) = v_{Soil}^{f-OC} \times 0.35\,L\ kg^{-1} \times K_{OW} \times \left(\frac{\rho_{OC}\left(kg\,m^{-3}\right)}{1000\,L\,m^{-3}} \right)$$

$$K_{Soil-W}(unitless) = 0.025 \times 0.35\,L\ kg^{-1} \times 10^{4.30} \times \left(\frac{1000\ kg\,m^{-3}}{1000\,L\,m^{-3}} \right)$$

$$K_{Soil-W}(unitless) = 1.7_5 \times 10^2$$

Since K_{Soil-W} is the ratio Z_{Soil}/Z_W, we can determine Z_W from the Henry's Law constant and use it to determine Z_{Soil}. However, note that the Henry's Law constant has pressure units of atmospheres, which is commonly encountered and requires a conversion factor of 101325 $Pa\ atm^{-1}$:

$$Z_W = \frac{1}{H}$$

$$Z_W = \frac{1}{2.00 \times 10^{-3} atm\ m^3\ mol^{-1} \times 101325\, Pa\, atm^{-1}}$$

$$Z_W = 4.93_5 \times 10^{-3}\ mol\ Pa^{-1} m^{-3}$$

Now, Z_{Soil} follows directly from K_{Soil-W}:

$$Z_{Soil} = K_{Soil-W} Z_W$$

$$Z_{Soil} = 1.7_5 \times 10^2 \times 4.93_5 \times 10^{-3}\ mol\ Pa^{-1} m^{-3}$$

$$Z_{Soil} = 8.6_2 \times 10^{-1}\ mol\ Pa^{-1} m^{-3}$$

Alternatively, we could determine Z_{Soil} directly from K_{OW} and Z_W:

$$Z_{Soil} = v_{Soil}^{f-OC} \times 0.35L\ kg^{-1} \times K_{OW} \times Z_W \times \left(\frac{\rho_{OC}\left(kg\, m^{-3}\right)}{1000\, L\, m^{-3}} \right)$$

$$Z_{Soil} = 0.025 \times 0.35L\ kg^{-1} \times 10^{4.30} \times 4.93_5 \times 10^{-3}\ mol\ Pa^{-1} m^{-3} \times \left(\frac{1000\, kg\, m^{-3}}{1000\, L\, m^{-3}} \right)$$

$$Z_{Soil} = 8.6_2 \times 10^{-1} mol\ Pa^{-1} m^{-3}$$

Worked Example 1.18

100 g of dry soil is mixed with 500 mL of water. To this mixture, 0.13 g of a chemical with molar mass = 200 $g\ mol^{-1}$ is added. Determine the concentrations and molar percentages of the chemical in the soil and water compartments, assuming insignificant water content in the soil itself. As in Worked Example 1.17, assume that the chemical has log K_{OW} = 4.30 and a Henry's Law constant $H = 2.00 \times 10^{-3}$ $atm\ m^3\ mol^{-1}$. Use a soil density of 3000 $kg\ m^{-3}$ and an organic carbon mass fraction $m_{Soil}^{f-OC} = 0.020$.

As before, the first task is to determine the various fugacity capacities for the compartments in question, in this case Z_{Soil} and Z_W. First, determine Z_W from the Henry's Law constant:

$$Z_W = \frac{1}{H}$$

$$Z_W = \frac{1}{2.00 \times 10^{-3} atm\ m^3\ mol^{-1} \times 101325\, Pa\, atm^{-1}}$$

$$Z_W = 4.93_5 \times 10^{-3}\ mol\ Pa^{-1} m^{-3}$$

Now determine Z_{Soil}:

$$Z_{Soil} = m_{Soil}^{f-OC} \times 0.35L\ kg^{-1} \times K_{OW} \times Z_W \times \left(\frac{\rho_{Soil}\left(kg\, m^{-3}\right)}{1000\, L\, m^{-3}} \right)$$

$$Z_{Soil} = 0.020 \times 0.35L\ kg^{-1} \times 10^{4.30} \times 4.93_5 \times 10^{-3}\ mol\ Pa^{-1} m^{-3} \times \left(\frac{3000\, kg\ m^{-3}}{1000\, L\, m^{-3}} \right)$$

$$Z_{Soil} = 2.0_7\ mol\, Pa^{-1} m^{-3}$$

To determine the system fugacity, we need these fugacity capacities as well as the compartment volumes. Since the quantity of soil is given as a mass, compute the volume of the soil compartment from its mass and density:

$$V_{Soil} = \frac{m_{Soil}}{\rho_{Soil}} = \frac{100.\,g \times 10^{-3}\,kg\,g^{-1}}{3000\,kg\;m^{-3}} = 3.33_3 \times 10^{-5}\,m^3$$

The system fugacity for a set of equilibrating compartments is given by:

$$f_{Sys} = \frac{m}{\sum_{i=1}^{j} Z_i\,V_i}$$

$$f_{Sys} = \frac{\left(\frac{mass}{M}\right)}{Z_{Soil}\,V_{Soil} + Z_W\,V_W}$$

$$f_{Sys} = \frac{\left(\frac{0.13\,g}{200.\,g\,mol^{-1}}\right)}{2.0_7\,mol\,Pa^{-1}m^{-3} \times 3.33_3 \times 10^{-5}\,m^3 + 4.93_5 \times 10^{-3}\,mol\;Pa^{-1}m^{-3} \times 500.\;mL \times 10^{-6}\,m^3\,mL^{-1}}$$

$$f_{Sys} = \frac{6.5_0 \times 10^{-4}\,mol}{7.1_4 \times 10^{-5}\,mol\,Pa^{-1}} = 9.1_1\,Pa$$

With the fugacity in hand, the concentrations are:

$$C_W = Z_W f_{Sys} = 4.93_5 \times 10^{-3}\,mol\;Pa^{-1}m^{-3} \times 9.1_1\,Pa = 4.4_9 \times 10^{-2}\,mol\,m^{-3}$$

$$C_{Soil} = Z_{Soil} f_{Sys} = 2.0_7\,mol\,Pa^{-1}m^{-3} \times 9.1_1\,Pa = 1.8_8 \times 10^{1}\,mol\,m^{-3}$$

The molar amounts in each compartment are given by:

$$m_W = C_W V_W = 4.4_9 \times 10^{-2}\,mol\,m^{-3} \times 500\,mL \times 10^{-6}\,m^3\,mL^{-1} = 2.2_5 \times 10^{-5}\,mol$$

$$m_{Soil} = C_{Soil} V_{Soil} = 1.8_8 \times 10^{1}\,mol\,m^{-3} \times 3.33_3 \times 10^{-5}\,m^3 = 6.2_8 \times 10^{-4}\,mol$$

As a check, do a mass balance:

$$m \stackrel{?}{=} m_W + m_{Soil}$$

$$\frac{0.13\,g}{200\,g\,mol^{-1}} \stackrel{?}{=} 2.2_5 \times 10^{-5}\,mol + 6.2_8 \times 10^{-4}\,mol$$

$$6.5 \times 10^{-4}\,mol = 6.5 \times 10^{-4}\,mol$$

Finally, the mole fractions as percentages:

$$\%x_W = \frac{m_W}{m} \times 100\% = \frac{2.2_5 \times 10^{-5}\,mol}{6.5 \times 10^{-4}\,mol} \times 100\% = 3.5\%$$

$$\%x_{Soil} = \frac{m_{Soil}}{m} \times 100\% = \frac{6.2_8 \times 10^{-4}\,mol}{6.5 \times 10^{-4}\,mol} \times 100\% = 97\%\left(96._5\%\right)$$

1.5 PARTITIONING TO BIOTA

Living organisms exist within environmental compartments with which they exchange chemicals and other matter through absorption, adsorption, ingestion, egestion and reproduction. In the simplest treatment of biota, the organisms are considered to be at equilibrium with the medium in which they exist. Under this assumption, we can consider the fugacity of any organism to be equal to that of the surrounding medium:

$$f_{Biota} = f_{Surroundings}$$

The only missing piece of the problem is the fugacity capacity of the organism. This may be estimated by making the assumption that, for most organic chemicals that partition predominantly to lipids, only the fatty or lipid-rich portions of the organism will uptake a chemical to any significant extent. Moreover, the further assumption may also be made that the volume of this "lipid fraction" can be represented by an equal volume of octanol. Under these assumptions, the fugacity capacity of the organism is given by:

$$\boxed{Z_{Biota} = v_{Biota}^{f-L} Z_O}$$

Here, v_{Biota}^{f-L} is the lipid fraction of the biota, the volume fraction of the total biota that is lipid-rich. Since we commonly have K_{OW} available to us, and almost always have to calculate Z_W for other reasons, we can more conveniently write this as:

$$\boxed{Z_{Biota} = v_{Biota}^{f-L} K_{OW} Z_W}$$

Worked Example 1.19

Determine the fugacity capacity of a chemical in a fish with a lipid volume fraction of 0.050, assuming the chemical has the following properties: log $K_{OW} = 4.20$, $H = 10.0\ Pa\ m^3\ mol^{-1}$.

Since we have the volume fraction of lipid in the fish and log K_{OW}, we need only determine the fugacity capacity of the chemical in water to be able to use the expression for Z_{Biota} given directly above.

The fugacity capacity of the chemical in water may be obtained directly from the Henry's Law constant:

$$Z_W = \frac{1}{H}$$

$$Z_W = \frac{1}{10.0\ Pa\ m^3\ mol^{-1}}$$

$$Z_W = 1.00 \times 10^{-1} mol\ Pa^{-1} m^{-3}$$

The fugacity capacity of the fish is now assumed to be given by the product of the fish lipid volume fraction and Z_O, the latter being replaced conveniently here by the product of K_{OW} and Z_W, as developed above:

$$Z_{Fish} = v_{Fish}^{f-L} Z_O = v_{Fish}^{f-L} K_{OW} Z_W$$

$$Z_{Fish} = 0.050 \times 10^{4.20} \times 1.00 \times 10^{-1} mol\ Pa^{-1} m^{-3}$$

$$Z_{Fish} = 7.9_2 \times 10^1 mol\ Pa^{-1} m^{-3}$$

Worked Example 1.20

A 25.0 L aquarium containing 100 fish of individual volume 1.0 cm^3 is treated with 15 g of the same chemical introduced in Worked Example 1.19. Determine the concentration and mole percentage of chemical in the fish and the water, again assuming the fish are 5.0% lipid by volume, and are at equifugacity with the water. Use a molar mass of 300 $g\ mol^{-1}$.

From the previous Worked Example, we know the fugacity capacities of both the water and the fish for this chemical:

$$Z_W = 1.00\times10^{-1}\ mol\ Pa^{-1}m^{-3}$$

$$Z_{Fish} = 7.9_2\times10^{1}\ mol\ Pa^{-1}m^{-3}$$

To calculate the system fugacity, we need to know the total fish volume and the resulting water volume:

$$V_{Fish} = 1.0\ cm^3 fish^{-1}\times10^{-6}m^3cm^{-3}\times100\,fish = 1.0_0\times10^{-4}m^3$$

$$V_W = V - V_{Fish} = 25.0\,L\times10^{-3}m^3L^{-1} - 1.0_0\times10^{-4}m^3 = 2.49\times10^{-2}m^3$$

With these values in hand, we can determine the system fugacity:

$$f_{Sys} = \frac{m}{\sum_{i=1}^{n} Z_i V_i}$$

$$f_{Sys} = \frac{\left(\frac{mass}{M}\right)}{Z_{Fish}V_{Fish} + Z_W V_W}$$

$$f_{Sys} = \frac{\left(\frac{15.g}{300.g\,mol^{-1}}\right)}{7.9_2\times10^{1}mol\ Pa^{-1}m^{-3}\times1.0_0\times10^{-4}m^3 + 1.00\times10^{-1}\ mol\ Pa^{-1}m^{-3}\times2.49\times10^{-2}m^3}$$

$$f_{Sys} = \frac{0.050\,mol}{1.04\times10^{-2}mol\ Pa^{-1}} = 4.8_0\,Pa$$

The concentrations in the fish and water follow directly from the products Zf:

$$C_{Fish} = Z_{Fish}f_{Sys} = 7.9_2\times10^{1}mol\ Pa^{-1}m^{-3}\times4.8_0\,Pa = 3.8_0\times10^{2}mol\,m^{-3}$$

$$C_W = Z_W f_{Sys} = 0.100\ mol\ Pa^{-1}m^{-3}\times4.8_0\,Pa = 4.8_0\times10^{-1}mol\,m^{-3}$$

The molar amounts are given by the products CV:

$$m_{Fish} = C_{Fish}V_{Fish} = 3.8_0\times10^{2}mol\,m^{-3}\times1.0_0\times10^{-4}m^3 = 3.8_0\times10^{-2}\,mol$$

$$m_W = C_W V_W = 4.8_0\times10^{-1}mol\,m^{-3}\times2.49\times10^{-2}m^3 = 1.2_0\times10^{-2}mol$$

As a check, let's do a mass balance:

$$m \stackrel{?}{=} m_{Fish} + m_W$$

$$\left(\frac{15.g}{300.g\,mol^{-1}}\right) \stackrel{?}{=} 3.8_0\times10^{-2}\,mol + 1.2_0\times10^{-2}mol$$

$$5.0\times10^{-2}mol = 5.0\times10^{-2}\,mol$$

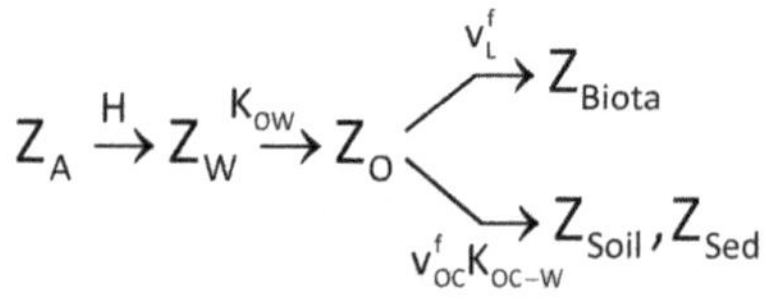

FIGURE 1.5 Illustration of the chain of relationships that connect essential fugacity capacities needed for environmental modelling, with the corresponding property needed to connect them.

Finally, the mole percentage of chemical in the fish and the water:

$$\%x_i = \left(\frac{C_i V_i}{m_i}\right) \times 100\%$$

$$\%x_{Fish} = \frac{3.8_0 \times 10^{-2}\, mol}{0.050\, mol} \times 100\% = 76\%$$

$$\%x_W = \frac{1.20 \times 10^{-2} mol}{0.050\, mol} \times 100\% = 24\%$$

Notice that, despite its relatively small volume, most of the chemical is concentrated in the fish by several orders of magnitude. This "bioconcentration" is a result of the relatively lipid-rich nature of fish compared with the surrounding waters, and the hydrophobic nature of the chemical in question (reflected in the high log K_{OW} = 4.20).

From the development above, it is apparent that the various fugacity capacities of interest to environmental modellers are related by a small number of properties. Figure 1.5 summarizes some of these relationships and shows the corresponding property that connects pairs of Z-values.

2 Equilibrium Partitioning in Complex Media

Bulk Z-Values

In this chapter, partitioning to media that are complex, i.e., a heterogeneous mixture of two or more phases, is introduced. The concept of bulk fugacity capacity is introduced as a volume-fraction-weighted sum of individual phase fugacity capacities. The calculation of mass fractions based on concentration and volume is introduced, as is the conversion of mass to volume fraction units. The handling of aerosols is discussed, and several approaches to determining the fugacity capacity of aerosols, and then the bulk fugacity capacity of air containing aerosols, are introduced. Bulk fugacity capacities for suspended solids in water, based on the volume fraction of organic carbon, the density of organic carbon and the relationship between the octanol–water partition ratio and K_{OC-W}, the organic carbon–water partition ratio, are discussed. A similar approach is used for soils containing air, water, organic and mineral matter, and roots. Finally, sediment bulk fugacity capacities are derived, similar to those of soils but omitting air and roots.

2.1 HETEROGENEOUS MIXTURES AND DISPERSED PHASES

Often, individual environmental media comprise heterogeneous mixtures, such as air containing aerosols, water containing suspended solids, soils containing air and water, and soils, sediments or water containing biota. If all subcompartments *within* a given medium are at equilibrium and therefore at equifugacity (which may often be safely assumed to be the case), the general approach is to simplify the problem by first establishing an overall or bulk fugacity capacity for the medium, computed from the Z-values of the various subcompartments present. The distribution of the chemical *between* media may then be determined by treating one or more heterogeneous media as if they were a homogeneous compartment with a single bulk fugacity capacity. Only subsequently is the chemical's distribution over the subcompartments themselves determined.

The general form for the bulk fugacity capacity of a medium with n components is simply a volume-fraction-weighted sum of "n" Z-values:

$$\boxed{Z_{Medium}^{Bulk} = \sum_{i=1}^{n} v_{Medium}^{f-i} Z_i}$$

For example, the bulk fugacity capacity of a soil composed of air, water, organic matter, mineral matter, and plant roots would be expressed as:

$$Z_{Soil}^{Bulk} = v_{Soil}^{f-A} Z_A + v_{Soil}^{f-W} Z_W + v_{Soil}^{f-OM} Z_{OM} + v_{Soil}^{f-MM} Z_{MM} + v_{Soil}^{f-Roots} Z_{Roots}$$

Word of Warning! The notation for volume fractions can lead to curious constructions that appear to refer to the volume fraction of a phase "within itself". Examples include the "volume fraction of air in air" v_A^{f-A} or the "volume fraction of water in water" v_W^{f-W}. In all such cases, the notation is intended to reflect the volume fraction of a pure fluid such as air or water

DOI: 10.1201/9781003657170-2

within the bulk mixture of that fluid and its other volume-occupying contents. The awkwardness derives from the linguistic convention of describing most mixtures of water or air containing any amount and number of other constituents simply as "water" or "air", respectively.

With the bulk fugacity capacity in hand, the fugacity in that medium is simply given by the ratio of the nominal or bulk concentration of the medium in question divided by its bulk fugacity capacity:

$$\boxed{f_i = \frac{C_i^{Bulk}}{Z_i^{Bulk}}}$$

Worked Example 2.1

A pond (V = 400 m^3) with water containing suspended solids with a volume fraction v_W^{f-SS}= 2.0 × 10^{-4} is contaminated with 100 g of a chemical with molar mass 200 $g\ mol^{-1}$. The Henry's Law constant for the chemical is H = 5.0 × 10^{-4} $atm\ m^3\ mol^{-1}$, and log K_{OW} = 5.85. Calculate the bulk fugacity capacity for the water, assuming an organic carbon content of 4.0% by volume in the suspended solids, and a density of organic carbon of 1000 $kg\ m^{-3}$. Use the bulk fugacity capacity to determine the overall fugacity of the system, then determine the percent mole fraction of the chemical in both the water and the suspended solids.

To calculate Z_W^{Bulk} we need to know the fugacity capacity of the water Z_W and of the suspended solids Z_{SS}. The fugacity capacity of the water may be obtained from the Henry's Law constant, with appropriate conversion from atm to Pa:

$$Z_W = \frac{1}{H}$$

$$Z_W = \frac{1}{5.0\times10^{-4} atm\, m^3\, mol^{-1}\times101325 Pa\, atm^{-1}} = 1.9_7\times10^{-2} mol\, Pa^{-1} m^{-3}$$

The fugacity capacity of the suspended solids is determined in the same manner as was introduced above for soils, but using the volume fraction of organic carbon in the suspended solids rather than in soil, along with the density of organic carbon:

$$Z_{SS} = v_{SS}^{f-OC}\times0.35L\ kg^{-1}\times K_{OW}\times Z_W\times\left(\frac{\rho_{OC}\left(kg\, m^{-3}\right)}{1000\, L\, m^{-3}}\right)$$

$$Z_{SS} = 0.040\times0.35L\ kg^{-1}\times10^{5.85}\times1.9_7\times10^{-2} mol\, Pa^{-1} m^{-3}\times\left(\frac{1000\, kg\, m^{-3}}{1000\, L\, m^{-3}}\right)$$

$$Z_{SS} = 1.9_6\times10^{2} mol\, Pa^{-1} m^{-3}$$

The bulk fugacity capacity now follows from the volume-fraction-weighted sum of the component fugacity capacities, with the volume fraction of pure water in the water containing suspended solids v_W^{f-W} being determined by difference:

$$Z_{Medium}^{Bulk} = \sum_{i=1}^{n} v_{Medium}^{f-i} Z_i$$

$$Z_W^{Bulk} = v_W^{f-SS} Z_{SS} + v_W^{f-W} Z_W$$

$$Z_W^{Bulk} = 2.0\times10^{-4}\times1.9_6\times10^{2} mol\, Pa^{-1} m^{-3} + (1-2.0\times10^{-4})\times1.9_7\times10^{-2} mol\, Pa^{-1} m^{-3}$$

$$Z_W^{Bulk} = 3.9_1\times10^{-2} mol\, Pa^{-1} m^{-3} + 1.9_7\times10^{-2} mol\, Pa^{-1} m^{-3}$$

$$Z_W^{Bulk} = 5.8_9\times10^{-2} mol\, Pa^{-1} m^{-3}$$

The nominal or bulk concentration of the chemical in the water C_W^{Bulk} is given by the amount of chemical added and the pond volume:

$$C_W^{Bulk} = \frac{m}{V} = \frac{\left(\dfrac{100.\,g}{200.\,g\,mol^{-1}}\right)}{400.\ m^3} = 1.25_0 \times 10^{-3}\,mol\,m^{-3}$$

The overall system fugacity, equal to the fugacity of the water, is obtained from the nominal or bulk concentration of the chemical in the water and the bulk fugacity capacity:

$$f_{Sys} = f_W = \frac{C_W^{Bulk}}{Z_W^{Bulk}}$$

$$f_W = \frac{1.25_0 \times 10^{-3}\,mol\,m^{-3}}{5.8_9 \times 10^{-2} mol\,Pa^{-1}\,m^{-3}} = 2.1_2 \times 10^{-2}\,Pa$$

The concentration in each phase is given by $C=Zf$:

$$C_W = Z_W f_W = 1.9_7 \times 10^{-2}\,mol\,Pa^{-1}\,m^{-3} \times 2.1_2 \times 10^{-2}\,Pa = 4.1_9 \times 10^{-4}\,mol\,m^{-3}$$

$$C_{SS} = Z_{SS} f_W = 1.9_6 \times 10^{2} mol\,Pa^{-1}\,m^{-3} \times 2.1_2 \times 10^{-2}\,Pa = 4.1_5\,mol\,m^{-3}$$

The amounts in each phase are given by CV:

$$m_W = C_W V_W = 4.1_9 \times 10^{-4}\,mol\,m^{-3} \times (1 - 2.0 \times 10^{-4}) \times 400\,m^3 = 0.16_8\,mol$$

$$m_{SS} = C_{SS} V_{SS} = 4.1_5\,mol\,m^{-3} \times (2.0 \times 10^{-4}) \times 400\,m^3 = 0.33_2\,mol$$

In terms of percentage mole fractions, we have:

$$\%x_W = \frac{0.16_8\,mol}{0.33_2\,mol + 0.16_8\,mol} \times 100 = 33._5\%$$

$$\%x_{SS} = \frac{0.33_2\,mol}{0.33_2\,mol + 0.16_8\,mol} \times 100 = 66._5\%$$

Despite its relatively modest total volume, the majority of the chemical is in the suspended solids, reflecting the chemical's high log K_{OW} value.

When phase A is dispersed within phase B, the mass fraction of a chemical in each phase, expressed in a fugacity framework will be given by the following expression for the mass fraction of the dispersed phase. The corresponding expression for the solvating phase given by analogy:

$$m_{A+B}^{f-A} = \frac{m_A}{m_A + m_B} = \frac{C_A V_A}{(C_A V_A + C_B V_B)} = \frac{Z_A f_{Sys} V_A}{(Z_A f_{Sys} V_A + Z_B f_{Sys} V_B)} = \frac{Z_A V_A}{(Z_A V_A + Z_B V_B)}$$

$$= \frac{Z_A V_A}{\left(Z_A v_A^f V^T + Z_B v_B^f V^T\right)} = \frac{Z_A V_A}{\left(Z_A v_A^f + Z_B v_B^f\right) V^T} = \frac{Z_A V_A}{\left(v_A^f Z_A + v_B^f Z_B\right) V^T} = \frac{Z_A V_A}{Z_B^{Bulk} V^T}$$

$$\boxed{m_{A+B}^{f-A} = \frac{Z_A V_A}{Z_B^{Bulk} V^T}} \qquad \boxed{m_{A+B}^{f-B} = \frac{Z_B V_B}{Z_B^{Bulk} V^T}}$$

Worked Example 2.2

Determine the fraction of chemical in each phase from the previous Worked Example, using the mass fraction expressions above. From before, we have $Z_W = 1.9_7 \times 10^{-2}\ mol\ Pa^{-1}\ m^{-3}$, $Z_{SS} = 1.9_6 \times 10^2\ mol\ Pa^{-1}\ m^{-3}$, $Z_W^{Bulk} = 5.8_9 \times 10^{-2}\ mol\ Pa^{-1}\ m^{-3}$, the total volume $V_T = 400\ m^3$, and the volume fraction of suspended solids $v_{SS}^f = 2.0 \times 10^{-4}$.

The only quantities we need to calculate to use the volume fraction expressions are the volumes of suspended solid and the water components:

$$V_{SS} = v_W^{f-SS} \times V^T$$

$$V_{SS} = 2.0 \times 10^{-4} \times 400.\ m^3$$

$$V_{SS} = 8.0_0 \times 10^{-2} m^3$$

The volume of the water component follows from calculation of the remaining volume fraction multiplied by the total volume:

$$V_W = v_W^{f-W} \times V^T = \left(1 - v_W^{f-SS}\right) \times V^T$$

$$V_W = \left(1 - 2.0 \times 10^{-4}\right) \times 400.\ m^3$$

$$V_W = 4.0_0 \times 10^2\ m^3$$

Note that the calculated volume of water is the same as that of the total volume, since the volume of suspended solids is so small. In this case, we could have ignored it and simply assumed $V_W = V_T$.

Now, using the remaining data:

$$m_W^{f-SS} = \frac{Z_{SS} V_{SS}}{Z_W^{Bulk} V^T}$$

$$m_W^{f-SS} = \frac{1.9_6 \times 10^2 mol\ Pa^{-1} m^{-3} \times 8.0_0 \times 10^{-2}\ m^3}{5.8_9 \times 10^{-2} mol\ Pa^{-1} m^{-3} \times 400.\ m^3}$$

$$m_W^{f-SS} = 0.66$$

$$m_W^{f-W} = \frac{Z_W V_W}{Z_W^{Bulk} V^T}$$

$$m_W^{f-W} = \frac{1.9_7 \times 10^{-2}\ mol\ Pa^{-1} m^{-3} \times 4.0_0 \times 10^2\ m^3}{5.8_9 \times 10^{-2} mol\ Pa^{-1} m^{-3} \times 400.\ m^3}$$

$$m_W^{f-W} = 0.34$$

These values are in accord with the mole percentages calculated in the previous Worked Example, as required.

Biota can also be considered to be a dispersed phase, existing as a discrete entity in soil, sediment, water or even air. In this way, we can use the same expressions to determine the fraction of the chemical in biota as we do for sediment particles, aerosols, and other dispersed phases.

Worked Example 2.3

Recalculate the fractions of chemical in the fish and water from Worked Example 1.20 using the expressions involving Z_W^{Bulk} given above. Recall we determined $Z_{Fish} = 7.9_2 \times 10^1\ mol\ Pa^{-1}m^{-3}$, $Z_W = 1.00 \times 10^{-1}\ mol\ Pa^{-1}m^{-3}$, $V_{Fish} = 1.0_0 \times 10^{-4}\ m^3$ and $V_W = 2.49 \times 10^{-2}\ m^3$.

The bulk fugacity capacity is given by a volume-fraction-weighted sum of the component fugacity capacities. We first need to determine the volume fraction of fish:

$$v_W^{f-Fish} = \frac{V_{Fish}}{V^T} = \frac{V_{Fish}}{V_{Fish} + V_W}$$

$$v_W^{f-Fish} = \frac{1.0_0 \times 10^{-4}\ m^3}{1.0_0 \times 10^{-4}\ m^3 + 2.49 \times 10^{-2}\ m^3} = 4.0_0 \times 10^{-3}$$

Now calculate the bulk fugacity capacity for the water phase as a volume-fraction-weighted sum of the components:

$$Z_W^{Bulk} = \sum_{i=1}^{n} v_i^f Z_i = v_W^{f-Fish} Z_{Fish} + v_W^{f-W} Z_W$$

$$Z_W^{Bulk} = 4.0_0 \times 10^{-3} \times 7.9_2 \times 10^1\, mol\, Pa^{-1}\, m^{-3} + \left(1 - 4.0_0 \times 10^{-3}\right) \times 1.00 \times 10^{-1}\, mol\, Pa^{-1}\, m^{-3}$$

$$Z_W^{Bulk} = 4.1_6 \times 10^{-1}\, mol\, Pa^{-1}\, m^{-3}$$

The mass fractions of the chemical in the fish and the water are given by:

$$m_W^{f-Fish} = \frac{Z_{Fish} V_F}{Z_W^{Bulk} V^T}$$

$$m_W^{f-Fish} = \frac{7.9_2 \times 10^1 mol\ Pa^{-1} m^{-3} \times 1.0_0 \times 10^{-4} m^3}{4.1_6 \times 10^{-1} mol\ Pa^{-1} m^{-3} \times 2.50 \times 10^{-2} m^3}$$

$$m_W^{f-Fish} = 0.76$$

$$m_W^{f-W} = \frac{Z_W V_W}{Z_W^{Bulk} V^T}$$

$$m_W^{f-W} = \frac{1.00 \times 10^{-1} mol\ Pa^{-1} m^{-3} \times 2.49 \times 10^{-2} m^3}{4.1_6 \times 10^{-1} mol\ Pa^{-1} m^{-3} \times 2.50 \times 10^{-2} m^3}$$

$$m_W^{f-W} = 0.24$$

These values are the same as the mole fractions determined in Worked Example 1.20, as required.

2.2 AEROSOLS

Aerosols are a special category of dispersed phase, consisting of small droplets of material suspended in air. A great deal of work has been done in attempt to estimate aerosol–air partition ratios, and several approaches have been proposed, as outlined subsequently. The approach to aerosols is similar to that for other dispersed phases, in that an aerosol fugacity capacity may be determined from the aerosol–air partition ratio:

$$Z_Q = K_{QA} \times Z_A$$

The challenge lies in determining K_{QA}. Aerosol–air partitioning is traditionally reported as the gas–particle partition ratio K_P. This ratio has units of $m^3\ \mu g^{-1}$ since chemical concentrations in aerosols, collected on a filter, are typically reported in units of *ng* chemical per μg of aerosol, whereas air concentrations are reported in units of *ng* of chemical per m^3 of air:

$$K_P\left(m^3 \mu g^{-1}\right) = \frac{C_Q\left(ng\, \mu g^{-1}\right)}{C_A\left(ng\, m^{-3}\right)}$$

K_P and the *unitless* K_{QA} are related via the density of the aerosols, ρ_Q:

$$K_{QA}(unitless) = \frac{C_Q\left(ng\, m^{-3}\right)}{C_A\left(ng\, m^{-3}\right)}$$

$$= \left[\frac{C_Q\left(ng\, \mu g^{-1}\right)}{C_A\left(ng\, m^{-3}\right)}\right] \times 10^9\, \mu g\, kg^{-1} \times \rho_Q\left(kg\, m^{-3}\right)$$

$$= K_P\left(m^3 \mu g^{-1}\right) \times 10^9\, \mu g\, kg^{-1} \times \rho_Q\left(kg\, m^{-3}\right)$$

Use of an assumed value of $\rho_Q = 2000\ kg\ m^{-3}$ leads to the following conversions:

$$K_{QA}\left(unitless\right) = K_P\left(m^3 \mu g^{-1}\right) \times 2 \times 10^{12}\ \mu g\ m^{-3}$$

or

$$\log K_{QA} = \log\left(K_P / m^3 \mu g^{-1}\right) + 12.30$$

Use of this relation required the relatively difficult-to-obtain experimental measurement of K_P for any chemical of interest. Therefore, it is of significant value to have an expression that relates the desired quantity K_{QA} to a more commonly available and theoretically estimable physico-chemical property of the chemical itself, rather than K_P.

There are several approaches to estimating K_{QA}, the first proposed by Mackay et al. (1986), and given by the following equation:

$$\boxed{K_{QA} = \frac{Z_Q}{Z_A} \simeq \frac{6 \times 10^6\, Pa}{P_A^{Sat}\,(Pa)}}$$

Here, the liquid or subcooled liquid vapour pressure P_A^{Sat} is given in *Pa*. The numerator 6×10^{-6} *Pa* is a fitted constant that may vary with the nature of the particulate matter, within the range 10^6 to 10^7. This corresponds to an uncertainty in log K_{QA} of about 0.5 log units. Thus, any calculation involving this expression has a fairly wide margin of uncertainty. This approach is now considered obsolete and has been mostly superseded by methods based on other more accurate approaches, presented below.

Harner and Bidleman (1998) proposed a correlation between K_P and K_{OA}:

$$\log\left(K_P / m^3 \mu g^{-1}\right) = \log K_{OA} + \log m_Q^{f-OM} - 11.91$$

Combining this expression with the previously derived expression for log K_{QA}, we can obtain a second equation for estimating K_{QA} from K_{OA}:

$$\log K_{QA} = \log K_{OA} + \log m_Q^{f-OM} - 11.91 + 12.30$$

$$\log K_{QA} = \log K_{OA} + \log m_Q^{f-OM} + 0.39$$

The corresponding expressions for K_{QA} and Z_Q follow:

$$K_{QA} = K_{OA} \times m_Q^{f-OM} \times 2.45$$

$$Z_Q = K_{OA} \times Z_A \times m_Q^{f-OM} \times 2.45$$

In terms of the fugacity capacity in octanol $Z_O = K_{OA} Z_A$, we have:

$$Z_Q = Z_O \times m_Q^{f-OM} \times 2.45$$

These expressions are of the same form as those given in Parnis and Mackay (2021). In that, the same assumption is made that all organic matter "acts" like octanol, such that the aerosol-air partition ratio is merely a volume-fraction-weighted octanol–air partition ratio:

$$K_{QA} = v_Q^{f-OM} K_{OA}$$

To express the organic matter in terms of the more commonly measured organic matter mass fraction requires a conversion involving the densities of organic matter and of the aerosols:

$$v_Q^{f-OM} = \frac{V_{OM}}{V_Q} = \frac{mass_{OM}(kg)/\rho_{OM}\left(kg\,m^{-3}\right)}{mass_Q(kg)/\rho_Q\left(kg\,m^{-3}\right)} = m_Q^{f-OM}\left(\frac{\rho_Q\left(kg\,m^{-3}\right)}{\rho_{OM}\left(kg\,m^{-3}\right)}\right)$$

Combining these last two equations yields a relation between *unitless* K_{OA} and *unitless* K_{QA}:

$$\boxed{K_{QA} = m_Q^{f-OM} K_{OA}\left(\frac{\rho_Q\left(kg\,m^{-3}\right)}{\rho_{OM}\left(kg\,m^{-3}\right)}\right)}$$

If the density of the organic matter is assumed to be the same as octanol (820 $kg\ m^{-3}$), and the density of the aerosols is taken as 2000 $kg\ m^{-3}$, we arrive at essentially the same result as from the derivation of Z_Q from the work of Harner and Bidleman:

$$Z_Q = m_Q^{f-OM} \times K_{OA} \times Z_A \times \left(\frac{2000\,kg\,m^{-3}}{820\,kg\,m^{-3}}\right)$$

$$Z_Q = m_Q^{f-OM} \times K_{OA} \times Z_A \times 2.44$$

$$Z_Q = m_Q^{f-OM} \times Z_O \times 2.44$$

Worked Example 2.4

A chemical has a log K_{OA} = 9.06. If the density of aerosols is 2000 $kg\ m^{-3}$ and its organic matter mass fraction is 0.50, what are the values of K_{QA} and Z_Q using the Parnis and Mackay expression based on K_{OA} and the organic matter mass fraction above? Assume the density of organic matter is the same as octanol, 820 $kg\ m^{-3}$, and $Z_A = 4.103 \times 10^{-4}$ $mol\ Pa^{-1}m^{-3}$.

We can directly estimate K_{QA} from the mass fraction of organic matter in the aerosol, K_{OA} and the densities of the aerosol and the organic matter:

$$K_{QA} = m_Q^{f-OM} K_{OA}\left(\frac{\rho_Q\left(kg\,m^{-3}\right)}{\rho_{OM}\left(kg\,m^{-3}\right)}\right)$$

$$K_{QA} = 0.50 \times 10^{9.06} \times \left(\frac{2000\,kg\,m^{-3}}{820\,kg\,m^{-3}}\right)$$

$$K_{QA} = 1.4_0 \times 10^9$$

The aerosol fugacity capacity is:

$$Z_Q = K_{QA} \times Z_A$$

$$Z_Q = 1.4_0 \times 10^9 \times 4.103 \times 10^{-4}\,mol\,Pa^{-1}\,m^{-3}$$

$$Z_Q = 5.7_4 \times 10^5\,mol\,Pa^{-1}\,m^{-3}$$

Calculation of the bulk fugacity capacity of air containing aerosols requires the volume fraction of aerosols, which must be determined from the aerosol concentration and density. Aerosol concentrations are typically measured in mass of aerosol per volume of air (e.g. $ng\ m^{-3}$), so unit conversion from mass fraction to volume fraction using the aerosol density is required.

For 1.0 m^3 of air containing aerosols at concentration C_A^Q($ng\ m^{-3}$) and density ρ_Q ($kg\ m^{-3}$), we have:

$$v_A^{f-Q} = \frac{V_Q}{V_T} = \frac{(mass(Q)/density(Q))}{V_T}$$

$$v_A^{f-Q} = \frac{\left[\left(C_A^Q\left(ng\,m^{-3}\,(air)\right)\times 1\,m^3\,(air)\right)\times 10^{-12}\,kg\,ng^{-1}\right] / \left[\rho_Q\left(kg\ m^{-3}\right)\right]}{1\,m^3\,(air)}$$

$$\boxed{v_A^{f-Q} = \frac{C_A^Q\left(ng\,m^{-3}\right)\times 10^{-12}\,kg\,ng^{-1}}{\rho_Q\left(kg\ m^{-3}\right)}}$$

Typical values of $3.0\times 10^4\ ng\ m^{-3}$ for C_A^Q and 1500 $kg\ m^{-3}$ for ρ_Q result in a corresponding aerosol volume fraction of 2.0×10^{-11}.

Worked Example 2.5

Determine the volume fraction of aerosols in an air sample with an aerosol particle concentration of 100 $\mu g\ m^{-3}$. Assume an aerosol density of 2000 $kg\ m^{-3}$.

First, note that the concentration of suspended particles (aerosols) is given in units of $\mu g\ m^{-3}$. Therefore, we need to convert it to the anticipated units of $ng\ m^{-3}$:

$$C_A^Q\left(ng\,m^{-3}\right) = C_A^Q\left(\mu g\,m^{-3}\right)\times 10^3\,ng\,\mu g^{-1}$$

$$C_A^Q = 100.\,\mu g\,m^{-3}\times 10^3\,ng\,\mu g^{-1} = 1.00\times 10^5\,ng\,m^{-3}$$

The volume fraction of aerosols is obtained directly as:

$$v_A^{f-Q} = \frac{C_A^Q\left(ng\,m^{-3}\right)\times 10^{-12}\,kg\,ng^{-1}}{\rho_Q\left(kg\ m^{-3}\right)}$$

$$v_A^{f-Q} = \frac{1.00\times 10^5\,ng\ m^{-3}\times 10^{-12}\,kg\,ng^{-1}}{2000\,kg\,m^{-3}}$$

$$v_A^{f-Q} = 5.00\times 10^{-11}$$

With the volume fraction in hand, we can calculate a bulk air fugacity capacity, according to the volume-fraction-weighted sum introduced earlier:

$$Z_A^{Bulk} = v_A^{f-Q} Z_Q + v_A^{f-A} Z_A$$

This relationship is illustrated in Figure 2.1.

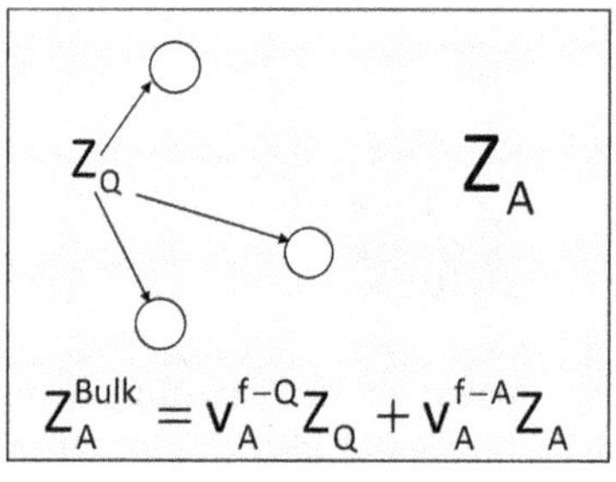

FIGURE 2.1 A graphical representation of the bulk fugacity calculation for air containing suspended aerosols (white circles) and their corresponding fugacity capacities.

Worked Example 2.6

Use $v_A^{f-Q} = 5.00 \times 10^{-11}$, $Z_A = 4.103 \times 10^{-4}\ mol\ Pa^{-1}\ m^{-3}$, and $Z_Q = 5.7_4 \times 10^5\ mol\ Pa^{-1} m^{-3}$ to calculate the bulk fugacity capacity of air containing aerosols.

The calculation is straightforward:

$$Z_A^{Bulk} = v_A^{f-Q} Z_Q + v_A^{f-A} Z_A$$

$$Z_A^{Bulk} = 5.00 \times 10^{-11} \times 5.7_4 \times 10^5\ mol\ Pa^{-1}\ m^{-3} + \left(1 - 5.00 \times 10^{-11}\right) \times 4.103 \times 10^{-4}\ mol\ Pa^{-1}\ m^{-3}$$

$$Z_A^{Bulk} = 4.39_0 \times 10^{-4}\ mol\ Pa^{-1}\ m^{-3}$$

Worked Example 2.7

Given the prevailing fugacity in the air as $f_A = 3.87 \times 10^{-8}\ Pa$, determine the concentration of the chemical in the bulk air, the free air, and the aerosols for the chemical in Worked Examples 2.4–2.6.

We can first use the previously calculated bulk fugacity capacity to determine the bulk concentration in air:

$$C_A^{Bulk} = Z_A^{Bulk} f_A$$

$$C_A^{Bulk} = 4.39_0 \times 10^{-4} mol\ Pa^{-1}\ m^{-3} \times 3.87 \times 10^{-8}\ Pa$$

$$C_A^{Bulk} = 1.70 \times 10^{-11} mol\ m^{-3}$$

The free air concentration is given by:

$$C_A = Z_A f_A$$

$$C_A = 4.103 \times 10^{-4} mol\ Pa^{-1}\ m^{-3} \times 3.87 \times 10^{-8}\ Pa$$

$$C_A = 1.59 \times 10^{-11} mol\ m^{-3}$$

The aerosol concentration is given by:

$$C_Q = Z_Q f_A$$

$$C_Q = 5.7_4 \times 10^5 mol\ Pa^{-1}\ m^{-3} \times 3.87 \times 10^{-8}\ Pa$$

$$C_Q = 2.2_2 \times 10^{-2} mol\ m^{-3}$$

It is often of interest to express the concentration in the aerosol in terms of amount per m^3 of the medium in which it is entrained, instead of per m^3 of aerosol, as a consequence of the common practice of sampling by filtration from the air itself. In this case:

$$C_A^Q = C_Q \times v_A^{f-Q}$$

Worked Example 2.8

Calculate the concentration of chemical from Worked Example 2.7 in the aerosol phase per m^3 of medium.

$$C_A^Q = C_Q \times v_A^{f-Q}$$

$$C_A^Q = 2.2_2 \times 10^{-2} mol\ m^{-3} (aerosol) \times 5.0_0 \times 10^{-11} \left(m^3\ aerosol / m^3\ bulk\ air\right)$$

$$C_A^Q = 1.1 \times 10^{-12} mol\ m^{-3} (bulk\ air)$$

Once expressed this way, we can see that the fraction of chemical in the aerosol phase is relatively small (about 6.5%), despite the fact that the relative concentration in the aerosol itself is extremely high (about 10^9 times C_A)! It is only by virtue of the small *relative* volume of aerosols that the chemical is not virtually completely transferred to the aerosol phase. The high concentration in the aerosols is a reflection of the chemical's high log K_{OA} value.

2.3 SUSPENDED SOLIDS IN WATER

Particulate matter in water is treated in a manner similar to what is used for soils, in that it is assumed to contain a volume fraction of organic carbon which "acts like" octanol from a partitioning point of view. The "total suspended particles" or "suspended solids" fugacity capacity is determined in terms of K_{OC-W}:

$$Z_{SS} = v_{SS}^{f-OC} \times 0.35\,L\;kg^{-1} \times K_{OW} \times Z_W \times \left(\frac{\rho_{OC}\left(kg\,m^{-3}\right)}{1000\,L\,m^{-3}} \right)$$

Here, we again have invoked the Seth relation between K_{OC-W} and K_{OW}, and the term in parentheses is due to the conversion of mass of suspended solids to corresponding volume.

The bulk fugacity capacity for water containing particulate matter is given as a volume-fraction-weighted sum of the components, in this case free water and suspended solids.

$$Z_W^{Bulk} = v_W^{f-W} Z_W + v_W^{f-SS} Z_{SS}$$

This relationship is shown schematically in Figure 2.2.

Worked Example 2.9

A litre of water containing particulate matter at equilibrium has a dissolved chemical concentration of 1.0×10^{-9} $mol\ m^{-3}$. The particulate matter has an associated volume fraction of 5.0×10^{-6}. The chemical has a log $K_{OW} = 6.00$ and a Henry's Law constant $H = 50\ Pa\ m^3\ mol^{-1}$. Assuming that the particulate matter has a volume fraction of organic carbon = 0.50 with an organic carbon density of 1000 $kg\ m^{-3}$, determine the concentration of the chemical in the particulate matter.

We need the fugacity capacities of water and the particulate matter. Z_W is obtained from the Henry's Law constant:

$$Z_W = \frac{1}{H}$$

$$Z_W = \frac{1}{50.\,Pa\,m^3\,mol^{-1}} = 2.0 \times 10^{-2}\,mol\,Pa^{-1}\,m^{-3}$$

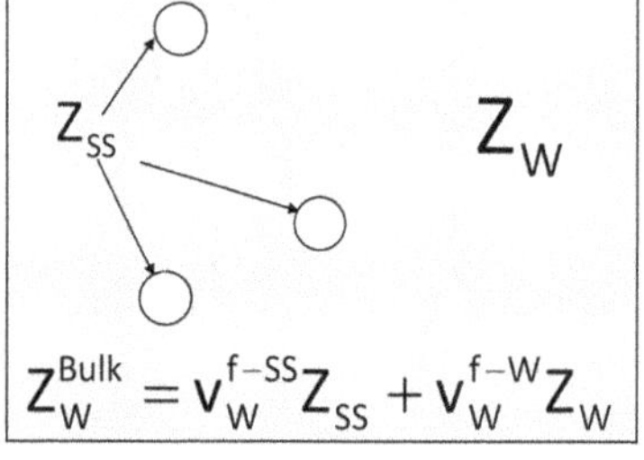

FIGURE 2.2 A graphical representation of the bulk fugacity calculation for water containing suspended solids (white circles), and their corresponding fugacity capacities.

The fugacity capacity of the particulate matter is given by:

$$Z_{SS} = v_{SS}^{f-OC} \times 0.35 L\ kg^{-1} \times K_{OW} \times Z_W \times \left(\frac{\rho_{OC}\left(kg\, m^{-3}\right)}{1000\, L\, m^{-3}} \right)$$

$$Z_{SS} = 0.50 \times 0.35 L\ kg^{-1} \times 10^{6.00} \times 2.0 \times 10^{-2}\, mol\, Pa^{-1}\, m^{-3} \times \left(\frac{1000\, kg\, m^{-3}}{1000\, L\, m^{-3}} \right)$$

$$Z_{SS} = 3.5_0 \times 10^3\, mol\, Pa^{-1}\, m^{-3}$$

Since the system is at equilibrium, we can calculate the fugacity from the free water concentration:

$$f_{Sys} = f_W = \frac{C_W}{Z_W}$$

$$f_{Sys} = \frac{1.0 \times 10^{-9} mol\ m^{-3}}{2.0 \times 10^{-2}\, mol\, Pa^{-1}\, m^{-3}} = 5.0 \times 10^{-8}\, Pa$$

The concentration of the chemical in the particulate matter is given by Zf:

$$C_{SS} = Z_{SS} f_{Sys}$$

$$C_{SS} = 3.5_0 \times 10^3\, mol\, Pa^{-1}\, m^{-3} \times 5.0 \times 10^{-8}\, Pa$$

$$C_{SS} = 1.7_5 \times 10^{-4}\, mol\, m^{-3}$$

The concentration in terms of water volume is given by:

$$C_W^{SS} = v_W^{f-SS} \times C_{SS}$$

$$C_W^{SS} = 5.0 \times 10^{-6} \times 1.7_5 \times 10^{-4}\, mol\, m^{-3}$$

$$C_W^{SS} = 8.8 \times 10^{-10}\, mol\, m^{-3}$$

Despite the very small volume fraction of particulate matter (5×10^{-6}), there is nearly as much chemical in the particulate matter as in the water, by total water volume. This high affinity of the chemical for the organic carbon in particulate matter is a reflection of its high log K_{OW} value.

2.4 SOILS

Soils are a complex mixture of organic and mineral matter, both of which contain water and air inside their particle pores. In addition, soils may contain roots from trees, plants, crops and other vegetation, as well as biota such as worms and other organisms. The primary challenge in modelling soils at equilibrium with their surroundings lies in determining the soil's bulk fugacity capacity. As before, this bulk fugacity capacity is expressed as a volume-fraction-weighted sum of the individual component fugacity capacities (see Figure 2.3):

$$Z_{Medium}^{Bulk} = \sum_{i=1}^{n} v_{Medium}^{f-i} Z_i$$

$$Z_{Soil}^{Bulk} = v_{Soil}^{f-A} Z_A + v_{Soil}^{f-W} Z_W + v_{Soil}^{f-OM} Z_{OM} + v_{Soil}^{f-MM} Z_{MM} + v_{Soil}^{f-Roots} Z_{Roots}$$

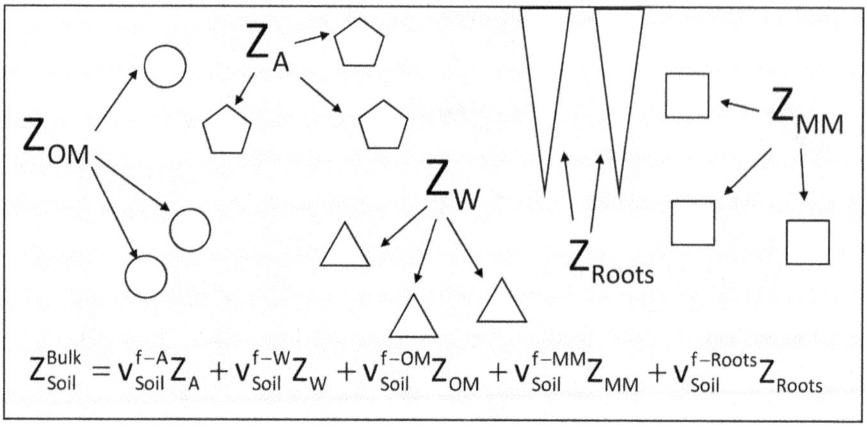

FIGURE 2.3 A graphical representation of the bulk fugacity calculation for soil containing air (pentagons), water (triangles), organic matter (circles), mineral matter (squares), and roots (elongated triangles) and their corresponding fugacity capacities.

The fugacity capacities of air and water in soil are as introduced above.

$$Z_A = \frac{1}{RT}$$

$$Z_W = \frac{Z_A}{K_{AW}} = \frac{1}{H}$$

The fugacity capacity of organic matter is proportional to that of organic carbon, with the volume fraction being the proportionality constant:

$$Z_{OM} = v_{OM}^{f\text{-}OC} Z_{OC}$$

To express this fugacity capacity in terms of mass fraction, we can relate the volume fraction to the mass fraction by the ratio of their densities:

$$v_{OM}^{f\text{-}OC} = \frac{V_{OC}}{V_{OM}} = \frac{\left(mass_{OC}(kg)/\rho_{OC}\left(kg\,m^{-3}\right)\right)}{\left(mass_{OM}(kg)/\rho_{OM}\left(kg\,m^{-3}\right)\right)} = m_{OM}^{f\text{-}OC}\left(\frac{\rho_{OM}\left(kg\,m^{-3}\right)}{\rho_{OC}\left(kg\,m^{-3}\right)}\right)$$

The fugacity capacity of organic matter expressed in terms of mass fraction follows by substitution:

$$Z_{OM} = m_{OM}^{f\text{-}OC} Z_{OC}\left(\frac{\rho_{OM}}{\rho_{OC}}\right)$$

Here the mass fraction of organic carbon in organic matter is generally assumed to be 0.56 (Sposito, 1989).

The fugacity capacity of organic carbon can be obtained from $K_{OC\text{-}W}$:

$$Z_{OC} = K_{OC\text{-}W} Z_W$$

Now, substituting this expression for Z_{OC} into that for Z_{OM}:

$$Z_{OM} = m_{OM}^{f-OC} K_{OC-W} Z_W \left(\frac{\rho_{OM}}{\rho_{OC}} \right)$$

From earlier, we have the Seth expression for K_{OC-W} in terms of K_{OW}:

$$K_{OC-W}(unitless) = 0.35\,L\ kg^{-1} \times K_{OW} \times \left(\frac{\rho_{OC}\left(kg\,m^{-3}\right)}{1000\,L\,m^{-3}} \right)$$

Substituting this expression for K_{OC-W} into the Z_{OM} expression above gives:

$$Z_{OM} = m_{OM}^{f-OC} \times \left[0.35\,L\ kg^{-1} \times K_{OW} \times \left(\frac{\rho_{OC}\left(kg\,m^{-3}\right)}{1000\,L\,m^{-3}} \right) \right] \times Z_W \times \left(\frac{\rho_{OM}}{\rho_{OC}} \right)$$

Cancellation of the density of organic carbon leads to the final expression, which was also derived above:

$$\boxed{Z_{OM} = m_{OM}^{f-OC} \times 0.35\,L\ kg^{-1} \times K_{OW} \times Z_W \times \left(\frac{\rho_{OM}\left(kg\,m^{-3}\right)}{1000\,L\,m^{-3}} \right)}$$

The fugacity capacity of mineral matter in soils is expressed in terms of the mineral matter–water partition ratio $K_{MM\text{-}W}$. If this partition ratio is measured as the amount of chemical per *kg* of mineral matter over the amount of chemical per *L* of water, we again have units of $L\ kg^{-1}$, as with $K_{OC\text{-}W}$. To convert this expression to a *unitless* ratio, it must be multiplied by the density of mineral matter. Since the density is typically expressed in $kg\ m^{-3}$, a conversion of water volume units is needed:

$$K_{MM-W}(unitless) = K_{MM-W}\left(L\,kg^{-1}\right) \times \left(\frac{\rho_{MM}\left(kg\,m^{-3}\right)}{1000\,L\,m^{-3}} \right)$$

Multiplying both sides by Z_W, we obtain the following expression for Z_{MM}:

$$\boxed{Z_{MM} = K_{MM-W}\left(L\,kg^{-1}\right) \times Z_W \times \left(\frac{\rho_{MM}\left(kg\,m^{-3}\right)}{1000\,L\,m^{-3}} \right)}$$

Finally, the fugacity capacity of plant roots is estimated in a similar way to biota, in terms of the volume fraction of lipid which is assumed to have the same fugacity capacity as octanol:

$$\boxed{Z_{Roots} = v_{Roots}^{f-L} \times Z_O = v_{Roots}^{f-L} \times K_{OW} Z_W}$$

The bulk fugacity capacity is calculated from the volume fractions of each component multiplied by its respective fugacity capacity. We generally have the necessary volume fractions for air, water, and roots, but not for organic and mineral matter. Instead, we often have the mass fraction of organic matter in dry soil from which the mass fraction of mineral matter can be deduced. The mass fractions of organic matter and mineral matter can then be related to their corresponding volume fractions in dry soil. Finally, these values can be converted into mass fractions in the soil as a whole, for the purpose of performing the bulk fugacity capacity calculation.

The relation between volume fraction of a component in a mixture of any two materials *A* and *B* and their mass fractions in the mixtures is as follows:

$$v_{A+B}^{f-A} = \frac{V_A}{V_A + V_B} = \left[\frac{1}{1+\left(\frac{V_B}{V_A}\right)}\right] = \left[\frac{1}{1+\left(\frac{mass_B / \rho_B}{mass_A / \rho_A}\right)}\right] = \left[\frac{1}{1+\left(\frac{(total\ mass)\times m_{A+B}^{f-B} / \rho_B}{(total\ mass)\times m_{A+B}^{f-A} / \rho_A}\right)}\right]$$

$$v_{A+B}^{f-A} = \left[\frac{1}{1+\left(\frac{m_{A+B}^{f-B} / \rho_B}{m_{A+B}^{f-A} / \rho_A}\right)}\right]$$

Thus, for dry soil deemed to be composed only of dry organic and mineral matter, we have for the volume fractions of organic and mineral matter:

$$v_{Dry\,Soil}^{f-OM} = \left[\frac{1}{1+\left(\frac{m_{DrySoil}^{f-MM} / \rho_{MM}}{m_{DrySoil}^{f-OM} / \rho_{OM}}\right)}\right]$$

$$v_{Dry\,Soil}^{f-MM} = \left[\frac{1}{1+\left(\frac{m_{DrySoil}^{f-OM} / \rho_{OM}}{m_{DrySoil}^{f-MM} / \rho_{MM}}\right)}\right]$$

The volume fractions of organic and mineral matter in the soil as a whole $v_{Soil}^{f-Dry\,Soil}$, with air, water, and roots included, are most conveniently given by the following difference:

$$v_{Soil}^{f-OM} + v_{Soil}^{f-MM} = v_{Soil}^{f-Dry\,Soil} = 1 - \left(v_{Soil}^{f-A} + v_{Soil}^{f-W} + v_{Soil}^{f-Roots}\right)$$

$$v_{Soil}^{f-OM} = \left[1 - \left(v_{Soil}^{f-A} + v_{Soil}^{f-W} + v_{Soil}^{f-Roots}\right)\right] \times v_{Dry\,Soil}^{f-OM}$$

$$v_{Soil}^{f-MM} = \left[1 - \left(v_{Soil}^{f-A} + v_{Soil}^{f-W} + v_{Soil}^{f-Roots}\right)\right] \times v_{Dry\,Soil}^{f-MM}$$

Worked Example 2.10

Estimate the bulk fugacity capacity of a soil with the following tabulated characteristics, for a chemical with $H = 2.29\ Pa\ m^3\ mol^{-1}$, log $K_{OW} = 6.19$ and $K_{MM-W} = 1.0\ L\ kg^{-1}$.

v_{Soil}^{f-A}	0.20
v_{Soil}^{f-W}	0.30
$v_{Soil}^{f-Roots}$	0.010
v_{Roots}^{f-L}	0.025
m_{OM}^{f-OC}	0.56
$m_{Dry\,Soil}^{f-OM}$	0.020
T	25.0°C
ρ_{Soil}, ρ_{MM}	2500 $kg\ m^{-3}$
ρ_{OC}, ρ_{OM}	1000 $kg\ m^{-3}$

The bulk fugacity capacity of the soil is given by the volume-fraction-weighted sum of the fugacity capacities of the soil components:

$$Z_{Soil}^{Bulk} = v_{Soil}^{f-A} Z_A + v_{Soil}^{f-W} Z_W + v_{Soil}^{f-OM} Z_{OM} + v_{Soil}^{f-MM} Z_{MM} + v_{Soil}^{f-Roots} Z_{Roots}$$

To use this expression, we first need to evaluate the various fugacity capacities and the volume fractions that are not already given.

The fugacity capacities are calculated as follows:

$$Z_A = \frac{1}{RT} = \frac{1}{8.3145\,Pa\,m^3\,mol^{-1}K^{-1} \times (25.0 + 273.15)K} = 4.034 \times 10^{-4}\,mol\,Pa^{-1}m^{-3}$$

$$Z_W = \frac{Z_A}{K_{AW}} = \frac{1}{H} = \frac{1}{2.29\,Pa\,m^3 mol^{-1}} = 4.36_7 \times 10^{-1} mol\,Pa^{-1}m^{-3}$$

$$Z_{OM} = m_{OM}^{f-OC} \times 0.35L\ kg^{-1} \times K_{OW} \times Z_W \times \left(\frac{\rho_{OM}\left(kg\,m^{-3}\right)}{1000\,L\,m^{-3}} \right)$$

$$Z_{OM} = 0.56 \times 0.35L\ kg^{-1} \times 10^{6.19} \times 4.36_7 \times 10^{-1} mol\,Pa^{-1}m^{-3} \times \left(\frac{1000\,kg\,m^{-3}}{1000\,L\,m^{-3}} \right)$$

$$Z_{OM} = 1.3_3 \times 10^5 mol\,Pa^{-1}m^{-3}$$

$$Z_{MM} = K_{MM-W}\left(L\,kg^{-1}\right) \times Z_W \times \left(\frac{\rho_{MM}\left(kg\,m^{-3}\right)}{1000\,L\,m^{-3}} \right)$$

$$Z_{MM} = 1.0\,L\,kg^{-1} \times 4.36_7 \times 10^{-1} mol\,Pa^{-1}m^{-3} \times \left(\frac{2500\,kg\,m^{-3}}{1000\,L\,m^{-3}} \right)$$

$$Z_{MM} = 1.0_9\ mol\,Pa^{-1}m^{-3}$$

$$Z_{Roots} = v_{Roots}^{f-L} \times Z_O = v_{Roots}^{f-L} \times K_{OW} \times Z_W$$

$$Z_{Roots} = 0.025 \times 10^{6.19} \times 4.36_7 \times 10^{-1} mol\,Pa^{-1}m^{-3}$$

$$Z_{Roots} = 1.6_9 \times 10^4\ mol\,Pa^{-1}m^{-3}$$

The volume fractions are listed below, either as given or calculated as follows:

$$v_{Soil}^{f-A} = 0.20$$

$$v_{Soil}^{f-W} = 0.30$$

$$v_{Soil}^{f-Roots} = 0.010$$

$$v_{Soil}^{f-Dry\,Soil} = \left[1 - \left(v_{Soil}^{f-A} + v_{Soil}^{f-W} + v_{Soil}^{f-Roots}\right)\right] = 0.49$$

The volume fractions of organic matter and mineral matter in soil require several steps to calculate. First, we need to determine the mass fraction of organic matter in dry soil, using the given mass fraction of organic carbon in the dry soil. The mass fractions of organic carbon and organic matter in dry soil are related as follows:

$$m_{OM}^{f-OC} = \frac{m_{DrySoil}^{f-OC}}{m_{DrySoil}^{f-OM}}$$

Rearrangement gives us the necessary conversion equation:

$$m_{DrySoil}^{f-OM} = \frac{m_{DrySoil}^{f-OC}}{m_{OM}^{f-OC}}$$

$$m_{DrySoil}^{f-OM} = \frac{0.020}{0.56} = 3.5_7 \times 10^{-2}$$

With this value in hand, we can calculate the volume fractions of organic and mineral matter in soil, according to the relations derived above, with the mass fraction of mineral matter determined by difference:

$$v_{Soil}^{f-OM} = v_{Soil}^{f-Dry\,Soil} \times v_{Dry\,Soil}^{f-OM} = v_{Soil}^{f-Dry\,Soil} \times \left[\frac{1}{1+\left(\frac{m_{DrySoil}^{f-MM}/\rho_{MM}}{m_{DrySoil}^{f-OM}/\rho_{OM}}\right)}\right]$$

$$v_{Soil}^{f-OM} = 0.49 \times \left[\frac{1}{1+\left(\frac{\left(1-3.5_7\times10^{-2}\right)/2500}{3.5_7\times10^{-2}/1000}\right)}\right] = 4.1_5 \times 10^{-2}$$

$$v_{Soil}^{f-MM} = v_{Soil}^{f-Dry\,Soil} \times \left[\frac{1}{1+\left(\frac{m_{DrySoil}^{f-OM}/\rho_{OM}}{m_{DrySoil}^{f-MM}/\rho_{MM}}\right)}\right]$$

$$v_{Soil}^{f-MM} = 0.49 \times \left[\frac{1}{1+\left(\frac{(0.02_0/0.56)/1000}{\left(1-(0.02_0/0.56)\right)/2500}\right)}\right] = 0.44_8$$

As a check, the sum of the organic matter and mineral matter volume fractions in the soil is:

$$v_{Soil}^{f-OM} + v_{Soil}^{f-MM} = 4.1_5 \times 10^{-2} + 0.44_8 \simeq 0.49$$

This is equal to the volume fraction of dry soil calculated above, as required by volume balance. Finally, the bulk fugacity capacity of the soil is given by:

$$Z_{Soil}^{Bulk} = v_{Soil}^{f-A} Z_A + v_{Soil}^{f-W} Z_W + v_{Soil}^{f-OM} Z_{OM} + v_{Soil}^{f-MM} Z_{MM} + v_{Soil}^{f-Roots} Z_{Roots}$$

$$\begin{aligned} Z_{Soil}^{Bulk} &= \left(0.20 \times 4.034 \times 10^{-4} mol\,Pa^{-1}m^{-3}\right) + \left(0.30 \times 4.36_7 \times 10^{-1} mol\,Pa^{-1}m^{-3}\right) \\ &\quad + \left(4.1_5 \times 10^{-2} \times 1.3_3 \times 10^{5} mol\,Pa^{-1}m^{-3}\right) + \left(0.44_8 \times 1.0_9\, mol\,Pa^{-1}m^{-3}\right) \\ &\quad + \left(0.010 \times 1.6_9 \times 10^{4}\, mol\,Pa^{-1}m^{-3}\right) \end{aligned}$$

$$Z_{Soil}^{Bulk} = 5.6_7 \times 10^{3}\, mol\,Pa^{-1}m^{-3}$$

Soils, considered as complex mixtures of air, water, organic matter, mineral matter, and roots offer an ideal scenario for illustration of the value of bulk fugacity capacity. In Worked Examples 2.11 and 2.12, we first consider soil interacting with a static air body, but only from the point of view of the air concentration at equilibrium. Here, only the bulk properties of the soil are needed, and the "details" of the chemical's distribution in the various components of the soil need not be considered. After this, having determined the fugacity of the chemical in the soil, we can use the fugacity capacities determined in Worked Example 2.10 to calculate the concentrations and mole percentages in each component of the soil.

Worked Example 2.11

3.00 *moles* of a chemical are homogeneously added to a soil body of volume $1.00 \times 10^3\ m^3$, which is in equilibrium contact with an air body of the same volume. The bulk fugacity capacity of the soil for this chemical is $5.6_7 \times 10^3\ mol\ Pa^{-1}\ m^{-3}$. Determine the concentration of the chemical in the air body.

Using the fugacity capacity of the air as calculated earlier: $4.034 \times 10^{-4}\ mol\ Pa^{-1}m^{-3}$.

The system fugacity is determined by:

$$f_{Sys} = \frac{m}{\sum_{i=1}^{j} Z_i V_i} = \frac{m}{Z_A V_A + Z_{Soil}^{Bulk} V_{Soil}}$$

$$f_{Sys} = \frac{3.00\,mol}{4.034 \times 10^{-4} mol\ Pa^{-1}m^{-3} \times 1.00 \times 10^3 m^3 + 5.6_7 \times 10^3\ mol\ Pa^{-1}m^{-3} \times 1.00 \times 10^3 m^3}$$

$$f_{Sys} = 5.2_9 \times 10^{-7}\,Pa$$

The air concentration is directly given by *Zf*:

$$C_A = Z_A f_{Sys} = 4.034 \times 10^{-4} mol\,Pa^{-1}m^{-3} \times 5.2_9 \times 10^{-7}\,Pa = 2.1_3 \times 10^{-10} mol\,m^{-3}$$

Worked Example 2.12

Use the fugacity of the chemical from Worked Example 2.11 to determine the concentrations and mole percentages in the bulk soil, and in each of its components, using the fugacity capacities calculated in Worked Example 2.10.

Each of the concentrations follows directly from the product *Zf*:

$$C_{Soil}^{Bulk} = Z_{Soil}^{Bulk} f_{Sys} = 5.6_7 \times 10^3\ mol\,Pa^{-1}m^{-3} \times 5.2_9 \times 10^{-7}\,Pa = 3.0_0 \times 10^{-3} mol\,m^{-3}$$

$$C_{Soil}^{A} = Z_A f_{Sys} = 4.034 \times 10^{-4} mol\,Pa^{-1}m^{-3} \times 5.2_9 \times 10^{-7}\,Pa = 2.1_3 \times 10^{-10} mol\,m^{-3}$$

$$C_{Soil}^{W} = Z_W f_{Sys} = 4.36_7 \times 10^{-1} mol\,Pa^{-1}m^{-3} \times 5.2_9 \times 10^{-7}\,Pa = 2.3_1 \times 10^{-7}\,mol\,m^{-3}$$

$$C_{Soil}^{OM} = Z_{OM} f_{Sys} = 1.3_3 \times 10^5 mol\,Pa^{-1}m^{-3} \times 5.2_9 \times 10^{-7}\,Pa = 7.0_1 \times 10^{-2} mol\,m^{-3}$$

$$C_{Soil}^{MM} = Z_{MM} f_{Sys} = 1.0_9\,mol\,Pa^{-1}m^{-3} \times 5.2_9 \times 10^{-7}\,Pa = 5.7_7 \times 10^{-7}\,mol\,m^{-3}$$

$$C_{Soil}^{Roots} = Z_{Roots} f_{Sys} = 1.6_9 \times 10^4\ mol\,Pa^{-1}m^{-3} \times 5.2_9 \times 10^{-7}\,Pa = 8.9_4 \times 10^{-3} mol\,m^{-3}$$

The molar amounts in each soil subcompartment, as well as the air above the soil, are given by:

$$m_{Soil}^{A} = C_{Soil}^{A} v_{Soil}^{f-A} V_{Soil} = 2.1_3 \times 10^{-10} mol\,m^{-3} \times 0.20 \times 1.00 \times 10^3 m^3 = 4.2_7 \times 10^{-8}\,mol$$

$$m_{Soil}^{W} = C_{Soil}^{W} v_{Soil}^{f-W} V_{Soil} = 2.3_1 \times 10^{-7}\,mol\,m^{-3} \times 0.30 \times 1.00 \times 10^3 m^3 = 6.9_3 \times 10^{-5} mol$$

$$m_{Soil}^{OM} = C_{Soil}^{OM} v_{Soil}^{f-OM} V_{Soil} = 7.0_1 \times 10^{-2} mol\,m^{-3} \times 4.1_5 \times 10^{-2} \times 1.00 \times 10^3 m^3 = 2.9_1 mol$$

$$m_{Soil}^{MM} = C_{Soil}^{MM} v_{Soil}^{f-MM} V_{Soil} = 5.7_7 \times 10^{-7}\,mol\,m^{-3} \times 0.44_8 \times 1.00 \times 10^3 m^3 = 2.5_9 \times 10^{-4} mol$$

$$m_{Soil}^{Roots} = C_{Soil}^{Roots} v_{Soil}^{f-Roots} V_{Soil} = 8.9_4 \times 10^{-3} mol\,m^{-3} \times 0.010 \times 1.00 \times 10^3 m^3 = 8.9_4 \times 10^{-2} mol$$

$$m_A^A = C_A V_A = 2.1_3 \times 10^{-10} mol\,m^{-3} \times 1.00 \times 10^3 m^3 = 2.1_3 \times 10^{-7}\,mol$$

As a check, the sum of these molar amounts plus the content of the air above the soil is 3.0 *mol*, which is the total moles of chemical added to the system.

Note that essentially all of the chemical is in the soil, as can also be shown using only the bulk concentration of the soil and the soil volume:

$$m_{Soil}^{Bulk} = C_{Soil}^{Bulk} V_{Soil} = 3.0_0 \times 10^{-3} mol\, m^{-3} \times 1.00 \times 10^3 m^3 = 3.0_0\, mol$$

The percent mole fractions are:

$$\%x_A = \left(\frac{m_A}{m}\right) \times 100\% = \frac{2.1_3 \times 10^{-7}\, mol}{3.0\, mol} = 7.1 \times 10^{-6}\,\%$$

$$\%x_{Soil}^{A} = \left(\frac{m_{Soil}^{A}}{m}\right) \times 100\% = \frac{4.2_7 \times 10^{-8}\, mol}{3.0\, mol} = 1.4 \times 10^{-6}\,\%$$

$$\%x_{Soil}^{W} = \left(\frac{m_{Soil}^{W}}{m}\right) \times 100\% = \frac{6.9_3 \times 10^{-5}\, mol}{3.0\, mol} = 2.3 \times 10^{-3}\,\%$$

$$\%x_{Soil}^{OM} = \left(\frac{m_{Soil}^{OM}}{m}\right) \times 100\% = \frac{2.9_1\, mol}{3.0\, mol} = 97\,\%$$

$$\%x_{Soil}^{MM} = \left(\frac{m_{Soil}^{MM}}{m}\right) \times 100\% = \frac{2.5_9 \times 10^{-4}\, mol}{3.0\, mol} = 8.6 \times 10^{-3}\,\%$$

$$\%x_{Soil}^{Roots} = \left(\frac{m_{Soil}^{Roots}}{m}\right) \times 100\% = \frac{8.9_4 \times 10^{-2}\, mol}{3.0\, mol} = 3.0\,\%$$

Here, it is clear that the high hydrophobicity of the chemical (log K_{OW} = 6.19) leads to its concentration in the "octanol-like" phases of organic matter and the roots.

2.5 SEDIMENTS

Sediments are similar to soils in that they are composed of water, organic matter, and mineral matter, but they differ in that they do not contain appreciable amounts of air. As with soils, the principal task is evaluating the fugacity capacities of these three components, and combining them into a bulk value using volume fractions:

$$Z_{Medium}^{Bulk} = \sum_{i=1}^{n} \mathrm{v}_{Medium}^{f-i} Z_i$$

$$Z_{Sed}^{Bulk} = \mathrm{v}_{Sed}^{f-OM} Z_{OM} + \mathrm{v}_{Sed}^{f-MM} Z_{MM} + \mathrm{v}_{Sed}^{f-W} Z_W$$

Figure 2.4 illustrates this relationship for sediments.

The fugacity capacity of water in soil is as introduced above:

$$Z_W = \frac{1}{H}$$

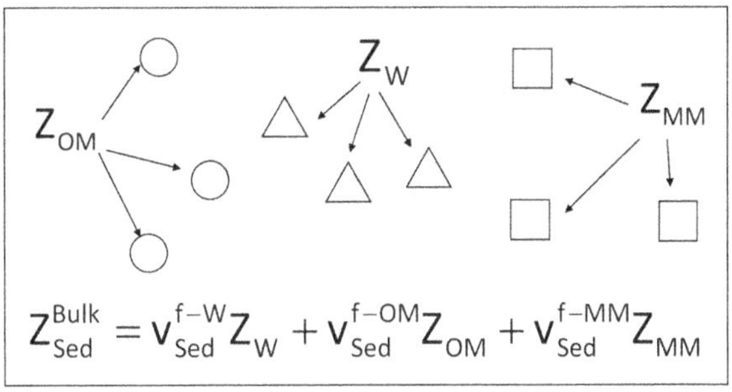

FIGURE 2.4 A graphical representation of the bulk fugacity calculation for sediment containing water (triangles), organic matter (circles), and mineral matter (squares) and their corresponding fugacity capacities.

The fugacity capacities for organic matter and mineral matter are the same as developed earlier:

$$Z_{OM} = m_{OM}^{f-OC} \times 0.35\,L\ kg^{-1} \times K_{OW} \times Z_W \times \left(\frac{\rho_{OM}\left(kg\,m^{-3}\right)}{1000\,L\,m^{-3}} \right)$$

$$Z_{MM} = K_{MM-W}\left(L\,kg^{-1}\right) \times Z_W \times \left(\frac{\rho_{MM}\left(kg\,m^{-3}\right)}{1000\,L\,m^{-3}} \right)$$

The volume fractions of organic and mineral matter are not usually available directly. Instead, as with soils, we typically have the mass fraction of organic carbon in the dry sediment, and the volume fraction of water. The volume fraction of dry matter, which is the sum of the organic and mineral matter volume fractions, is easily determined from the water volume fraction:

$$v_{Sed}^{f-DrySed} = v_{Sed}^{f-OM} + v_{Sed}^{f-MM} = 1 - v_{Sed}^{f-W}$$

Assuming the dry sediment only contains organic and mineral matter, the volume fractions of organic matter and mineral matter in the dry sediment are as developed earlier:

$$v_{Dry\,Sed}^{f-OM} = \left[\frac{1}{1 + \left(\dfrac{m_{Sed}^{f-MM} / \rho_{MM}}{m_{Sed}^{f-OM} / \rho_{OM}} \right)} \right]$$

$$v_{Dry\,Sed}^{f-MM} = \left[\frac{1}{1 + \left(\dfrac{m_{Sed}^{f-OM} / \rho_{OM}}{m_{Sed}^{f-MM} / \rho_{MM}} \right)} \right]$$

Finally, the individual volume fractions of organic and mineral matter in the sediment as a whole are given by the product of the volume fraction of dry sediment multiplied by the volume fraction of either the organic or mineral matter in the dry sediment.

$$v_{Sed}^{f-OM} = v_{Sed}^{f-DrySed} \times v_{Dry\,Sed}^{f-OM} = \left(1 - v_{Sed}^{f-W}\right) \times v_{Dry\,Sed}^{f-OM}$$

$$v_{Sed}^{f-MM} = v_{Sed}^{f-DrySed} \times v_{Dry\,Sed}^{f-MM} = \left(1 - v_{Sed}^{f-W}\right) \times v_{Dry\,Sed}^{f-MM}$$

Worked Example 2.13

Calculate the bulk fugacity capacity of a sediment containing 80% water by volume, a mass fraction of organic carbon of 0.050, a mass fraction of organic carbon in organic matter of 0.56 and $K_{MM-W} = 1.00\ L\ kg^{-1}$, for a chemical with $H = 300\ Pa\ m^3\ mol^{-1}$ and $\log K_{OW} = 6.00$. Use densities of 1000 and 2500 $kg\ m^{-3}$ for the density of organic and mineral matter, respectively.

To calculate the bulk fugacity capacity of the sediment, we need the fugacity capacities and volume fractions of the sediment water, organic matter and mineral matter.

The water fugacity capacity is:

$$Z_W = \frac{1}{H}$$

$$Z_W = \frac{1}{300.\,Pa\,m^3 mol^{-1}}$$

$$Z_W = 3.33_3 \times 10^{-3} mol\,Pa^{-1} m^{-3}$$

The organic matter and mineral matter fugacity capacities are:

$$Z_{OM} = m_{OM}^{f-OC} \times 0.35\,L\ kg^{-1} \times K_{OW} \times Z_W \times \left(\frac{\rho_{OM}\left(kg\,m^{-3}\right)}{1000\,L\,m^{-3}} \right)$$

$$Z_{OM} = 0.56 \times 0.35\,L\ kg^{-1} \times 10^{6.00} \times 3.33_3 \times 10^{-3} mol\,Pa^{-1} m^{-3} \times \left(\frac{1000\,kg\,m^{-3}}{1000\,L\,m^{-3}} \right)$$

$$Z_{OM} = 6.5_3 \times 10^2 mol\,Pa^{-1} m^{-3}$$

$$Z_{MM} = K_{MM-W}\left(L\,kg^{-1}\right) \times Z_W \times \left(\frac{\rho_{MM}\left(kg\,m^{-3}\right)}{1000\,L\,m^{-3}} \right)$$

$$Z_{MM} = 1.00\,L\,kg^{-1} \times 3.33_3 \times 10^{-3} mol\,Pa^{-1} m^{-3} \times \left(\frac{2500\,kg\,m^{-3}}{1000\,L\,m^{-3}} \right)$$

$$Z_{MM} = 8.33_3 \times 10^{-3} mol\,Pa^{-1} m^{-3}$$

The volume fraction of the water is given at 0.80. The volume fraction of the solid matter in the sediment is deduced by difference:

$$v_{Sed}^{f-W} = 0.80$$

$$v_{Sed}^{f-Dry\,Sed} = 1 - v_{Sed}^{f-W} = 1 - 0.80 = 0.20$$

Now, the organic matter content in the sediment is not given, but the organic carbon content is provided as a mass fraction. Using the provided mass fraction of 0.56 for organic carbon content in organic matter, the mass fraction of organic matter in the sediment is:

$$m_{Sed}^{f-OM} = \frac{m_{Sed}^{f-OC}}{m_{OM}^{f-OC}} = \frac{0.050}{0.56} = 0.089_3$$

The remaining mass fraction in the dry sediment must be mineral matter, so that:

$$m_{Sed}^{f-MM} = m_{Sed}^{f-Dry\,Sed} - m_{Sed}^{f-OM} = 1 - 0.089_3 = 0.910_7$$

To calculate the bulk fugacity capacity for the overall sediment, we need to know the volume fractions of organic matter and mineral matter. As before, we can relate the volume and mass fractions of these as:

$$v_{Dry\,Sed}^{f-OM} = \left[\frac{1}{1+\left(\frac{m_{Sed}^{f-MM}/\rho_{MM}}{m_{Sed}^{f-OM}/\rho_{OM}}\right)}\right] = \left[\frac{1}{1+\left(\frac{0.910_7/2500\,kg\,m^{-3}}{0.089_3/1000\,kg\,m^{-3}}\right)}\right] = 0.19_7$$

$$v_{Dry\,Sed}^{f-MM} = \left[\frac{1}{1+\left(\frac{m_{Sed}^{f-OM}/\rho_{OM}}{m_{Sed}^{f-MM}/\rho_{MM}}\right)}\right] = \left[\frac{1}{1+\left(\frac{0.089_3/1000\,kg\,m^{-3}}{0.910_7/2500\,kg\,m^{-3}}\right)}\right] = 0.803_1$$

Note that we could have saved a bit of work by recognizing that the two volume fractions of *OM* and *MM* must sum to 1.0, since there's nothing else assumed to be in the dry sediment. As such, the volume fraction of the mineral matter must equal 1 minus that of the organic matter:

$$v_{Dry\,Sed}^{f-MM} = 1 - v_{Dry\,Sed}^{f-OM} = 1 - 0.19_7 = 0.80_3$$

Finally, recognizing that the dry sediment has a volume fraction of 0.20 with respect to the total wet sediment volume, we have the volume fractions for *OM* and *MM* in the wet sediment as:

$$v_{Sed}^{f-OM} = 0.20 \times v_{Dry\,Sed}^{f-OM} = 0.20 \times 0.19_7 = 3.9_4 \times 10^{-2}$$
$$v_{Sed}^{f-MM} = 0.20 \times v_{Dry\,Sed}^{f-MM} = 0.20 \times 0.803_1 = 0.16_1$$

With these values in hand, the bulk fugacity capacity can now be calculated:

$$Z_{Sed}^{Bulk} = v_{Sed}^{f-OM} Z_{OM} + v_{Sed}^{f-MM} Z_{MM} + v_{Sed}^{f-W} Z_W$$
$$Z_{Sed}^{Bulk} = \left(3.9_4 \times 10^{-2} \times 6.5_3 \times 10^2 mol\,Pa^{-1}m^{-3}\right) + \left(0.16_1 \times 8.33_3 \times 10^{-3} mol\,Pa^{-1}m^{-3}\right)$$
$$+ \left(0.80 \times 3.33_3 \times 10^{-3} mol\,Pa^{-1}m^{-3}\right)$$
$$Z_{Sed}^{Bulk} = 2.5_7 \times 10^1 mol\,Pa^{-1}m^{-3} + 1.3_4 \times 10^{-3} mol\,Pa^{-1}m^{-3} + 2.66_7 \times 10^{-3} mol\,Pa^{-1}m^{-3}$$
$$Z_{Sed}^{Bulk} = 2.5_7 \times 10^1 mol\,Pa^{-1}m^{-3}$$

Note that the bulk fugacity capacity of the sediment is essentially that of the organic matter fraction in the sediment. This fact reflects the relatively high affinity of the chemical for the organic matter content of the sediment.

3 Open Systems at Steady State
Introducing D-Values

This chapter introduces the movement of chemicals by bulk transport called advection and the loss of chemicals by transformational degradation. Both of these bulk processes can be handled in similar manners by defining *D*-values, which are a form of fugacity rate constant. The advection rate is shown to be determined as a product of volume flow rate and concentration, which can also be expressed as the advection *D*-value multiplied by the prevailing fugacity. The determination of the system fugacity when advection and/or degradation are active requires a molar flow balance of inflow and outflow. Bulk advection of particulate matter is introduced and shown to require the volume fraction of the matter to modify the bulk flow value. Sediment burial is introduced as an advective process. Degradation rates as related to rate constants for first-order decay and the associated *D*-values are discussed. The combination of *D*-values for advection and degradation is discussed, followed by introduction of the concept of residence time.

3.1 ADVECTION

Most environmental systems involve some degree of transport of chemicals as a result of bulk flow within, into, or out of one or more media compartment, or within the system itself between subcompartments. This mechanism of chemical transport is called advection. In essence, the chemicals "hitch a ride" within various carriers. Systems that have advective inputs or outputs are termed "open" systems. Bulk flow advective transport may take the form of chemicals entrained in the water flowing into or out of a lake, or by inflow of contaminated air or sediment into an air body or river segment, respectively. Soil runoff and sediment burial are also examples of advection, as discussed further in this chapter.

Figure 3.1 shows the basic bulk flow advective processes for a water body in contact with air and sediment, including sediment burial as a form of bulk advective loss from the system due to sediment removal from the system. Note that under the continuously stirred tank reactor assumption, the outflow concentration of each compartment is equal to the concentration of that compartment itself, whereas the inflow concentrations depend on external conditions.

In our prior consideration of closed system at equilibrium, we used a mass balance equation approach as our basis for determining the system fugacity. As mentioned in Section 1.2, these are termed "Level I" systems or models in the Mackay terminology. For systems at equilibrium involving steady-state input and output flow, a mass transfer rate equation approach is used instead, and therefore we will need to consider expressions of transfer process rates in fugacity terms. Such systems at steady-state and equilibrium, and their associated models are termed "Level II".

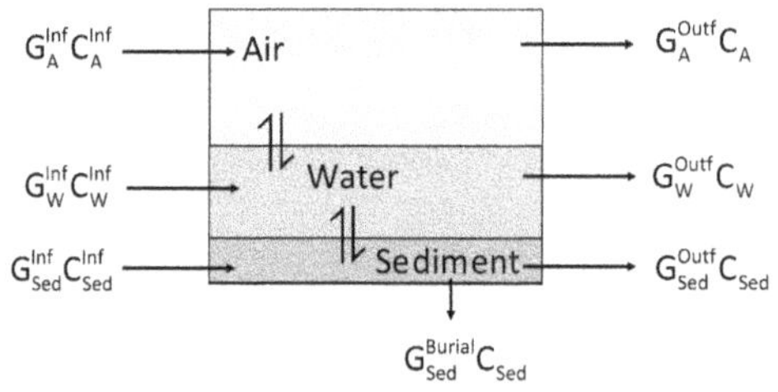

FIGURE 3.1 Schematic demonstrating advection into and out of an evaluative environment consisting of water in contact with air and sediment.

DOI: 10.1201/9781003657170-3

Word of Warning! The equilibrium and steady-state conditions are not the same. Equilibrium implies the thermodynamic condition that the rates of transport in both directions between any two compartments are equal, i.e., there is no net change in intercompartment transfer. The steady-state condition implies that the concentrations in all compartments are not changing with time. The latter condition may be met by systems NOT at equilibrium. Furthermore, systems at equilibrium may undergo concentration changes over time, yet still meet the condition of no net intercompartment transfer if the rate of achieving equilibrium is faster than the rate of concentration changes in all compartments.

The basic equation describing the chemical advection rate r ($mol\ h^{-1}$) associated with advective flow of a medium containing the chemical is given by the product of the medium flow rate G (m^3h^{-1}) and the concentration of chemical in that flowing medium C ($mol\ m^{-3}$):

$$\boxed{r_i = G_i \times C_i}$$

Worked Example 3.1

Calculate the advective flow rate for a chemical entrained in water flowing at a rate of G_W = 500 $m^3\ h^{-1}$ and containing a chemical at a concentration $C_W = 1.0 \times 10^{-3}\ mol\ m^{-3}$.

The flow rate is given by:

$$r = G_W \times C_W$$

$$r = 500.\ m^3h^{-1} \times 1.0 \times 10^{-3} mol\ m^{-3}$$

$$r = 5.0 \times 10^{-1} mol\ h^{-1}$$

For systems with more than one environmental compartment, the simplest scenario is one in which all compartments are at both steady state and at equilibrium. This means the amount of chemical in each compartment is not changing with time (steady-state condition) and the fugacity in all compartments is equal (equilibrium condition). Note that for a steady-state system with constant compartment volumes, the inflow and outflow rates for a given compartment must necessarily be equal. A convenient feature of this scenario is that it does not matter where any chemical emission occurs, since the entire model environment "instantly distributes" the added chemical among the various compartments in accord with the corresponding partition ratios.

For an open system at steady state, the total rate of chemical input I for "n" environmental compartments at equilibrium is given by:

$$\boxed{I = E + \sum_{i=1}^{n} G_i^{Inf} C_i^{Inf}}$$

Here, E is the emission rate at which a chemical is directly added to the environment by one or more processes in units of $mol\ h^{-1}$.

Word of Warning! The symbol "I" is also used in equilibrium calculations involving ionizing chemicals for the ratio of ionized to neutral forms and should not be confused with its use here for the total rate of chemical input into a system.

For an open system at steady state, the total inflow must equal the total outflow:

$$r^{Inf} = r^{Outf}$$

The outflow concentrations associated with any compartment will equal the concentration of that compartment itself, in which case we can write the following mass transfer rate balance for n compartments:

$$I = E + \sum_{i=1}^{n} G_i^{Inf} C_i^{Inf} = \sum_{i=1}^{n} G_i^{Outf} C_i$$

This relation can be re-expressed in terms of the system equilibrium fugacity as:

$$I = E + \sum_{i=1}^{n} G_i^{Inf} C_i^{Inf} = \sum_{i=1}^{n} G_i^{Outf} Z_i f_{Sys}$$

The rate of inflow and outflow for any compartment can be termed as that compartment's "advection rate", so that we can also write:

$$I = \sum_{i=1}^{n} G_i^{Adv} Z_i f_{Sys}$$

This expression leads directly to the expression for the system fugacity:

$$\boxed{f_{Sys} = \frac{I}{\sum_{i=1}^{n} G_i^{Adv} Z_i}}$$

The product GZ has units of $mol\ Pa^{-1}\ h^{-1}$, and is given the special symbol "D", a fugacity-based transport rate parameter:

$$\boxed{D_i^{Adv} = G_i^{Adv} Z_i}$$

In light of this definition, we can recast the expression for system fugacity as:

$$\boxed{f_{Sys} = \frac{I}{\sum_{i=1}^{n} D_i^{Adv}}}$$

Worked Example 3.2

A pond is fed by a stream with a flow rate of 5.0 $m^3\ h^{-1}$. The stream is contaminated with a chemical with $H = 50\ Pa\ m^3\ mol^{-1}$ at a concentration of $4.0 \times 10^{-1}\ mol\ m^{-3}$. There is also direct emission of the same chemical into the pond at a rate of $E = 10\ mol\ h^{-1}$. Assuming the pond is well-mixed, determine the outflow concentration of the chemical.

This is a relatively simple application of the previous multi-compartment development. In this case, the total rate of chemical input is given by:

$$I = E + G_W^{Inf} C_W^{Inf}$$

$$I = 10.\ mol\,h^{-1} + 5.0\,m^3h^{-1} \times 4.0 \times 10^{-1}\,mol\ m^{-3}$$

$$I = 12._0\ mol\,h^{-1}$$

There is only one D-value for advection, associated with the water:

$$D_W^{Adv} = G_W Z_W$$

The necessary fugacity capacity of water follows from the Henry's Law constant:

$$Z_W = \frac{1}{H}$$

$$Z_W = \frac{1}{50.\ Pa\ m^3 mol^{-1}} = 2.0_0 \times 10^{-2} mol\,Pa^{-1}\,m^{-3}$$

This gives a D-value of:

$$D_W^{Adv} = G_W Z_W$$

$$D_W^{Adv} = 5.0\ m^3h^{-1} \times 2.0_0 \times 10^{-2} mol\,Pa^{-1}\,m^{-3}$$

$$D_W^{Adv} = 1.0_0 \times 10^{-1}\ mol\,Pa^{-1}\,h^{-1}$$

The system fugacity is therefore:

$$f_{Sys} = \frac{I}{D_W^{Adv}}$$

$$f_{Sys} = \frac{12._0\ mol\,h^{-1}}{1.0_0 \times 10^{-1}\ mol\,Pa^{-1}\,h^{-1}} = 1.2_0 \times 10^2 Pa$$

The concentration in the water is:

$$C_W = Z_W f_{Sys}$$

$$C_W = 2.0_0 \times 10^{-2}\ mol\,Pa^{-1}\,m^{-3} \times 1.2_0 \times 10^2 Pa$$

$$C_W = 2.4\ mol\,m^{-3}$$

Worked Example 3.3

A pond of volume V_W = 500 m^3 is fed by a stream with a flow rate of 1.0 $m^3\ h^{-1}$. The stream is contaminated with a chemical with H = 50 $Pa\ m^3\ mol^{-1}$ at a concentration of $4.0 \times 10^{-2}\ mol\ m^{-3}$. There is also direct emission of the same chemical into the pond at a rate of $E = 2.0 \times 10^{-2}\ mol\ h^{-1}$. Above the pond is an air compartment of volume $V_A = 5.0 \times 10^3\ m^3$. Air flows into the air body at a rate of $1.0 \times 10^3\ m^3\ h^{-1}$ with a concentration of the chemical of $1.0 \times 10^{-4}\ mol\ m^{-3}$. Assuming the pond and air are both well-mixed, at steady state, and at equilibrium, determine the outflow concentration of chemical in the air and water, the molar outflow rates for each compartment and the amount of chemical in each compartment. We assume $Z_A = 4.0 \times 10^{-4}\ mol\ Pa^{-1}\ m^{-3}$ since no temperature is specified.

The total inflow rate I is given by the emission and advection rates:

$$I = E + G_W^{Inf} C_W^{Inf} + G_A^{Inf} C_A^{Inf}$$

$$I = 2.0 \times 10^{-2} mol\ h^{-1} + 1.0\ m^3 h^{-1} \times 4.0 \times 10^{-2} mol\ m^{-3} + 1.0 \times 10^3 m^3 h^{-1} \times 1.0 \times 10^{-4} mol\ m^{-3}$$

$$I = 2.0 \times 10^{-2} mol\ h^{-1} + 4.0 \times 10^{-2} mol\ h^{-1} + 1.0 \times 10^{-1} mol\ h^{-1}$$

$$I = 1.6_0 \times 10^{-1} mol\ h^{-1}$$

To determine the fugacity, we need to know the D-values for water and air compartment advection, which are products of the known G flow rates and the fugacity capacities Z_W and Z_A, respectively. Use $Z_A = 4.0 \times 10^{-4}\ mol\ Pa^{-1}\ m^{-3}$ and $Z_W = 2.0 \times 10^{-2}\ mol\ Pa^{-1}\ m^{-3}$.

The sum of advective D-values is given by:

$$\sum_{i=1}^{n} D_i^{Adv} = D_W^{Adv} + D_A^{Adv} = G_W^{Adv} Z_W + G_A^{Adv} Z_A$$

The individual D-values are:

$$D_W^{Adv} = 1.0\ m^3 h^{-1} \times 2.0 \times 10^{-2} mol\ Pa^{-1} m^{-3} = 2.0_0 \times 10^{-2} mol\ Pa^{-1} h^{-1}$$

$$D_A^{Adv} = 1.0 \times 10^3 m^3 h^{-1} \times 4.0 \times 10^{-4} mol\ Pa^{-1} m^{-3} = 4.0_0 \times 10^{-1} mol\ Pa^{-1} h^{-1}$$

The D-value sum is:

$$\sum_{i=1}^{n} D_i^{Adv} = 2.0 \times 10^{-2} mol\ Pa^{-1} h^{-1} + 4.0_0 \times 10^{-1} mol\ Pa^{-1} h^{-1}$$

$$\sum_{i=1}^{n} D_i^{Adv} = 4.2_0 \times 10^{-1} mol\ Pa^{-1} h^{-1}$$

Now, the system fugacity is:

$$f_{Sys} = \frac{I}{\sum_{i=1}^{n} D_i^{Adv}}$$

$$f_{Sys} = \frac{1.6_0 \times 10^{-1} mol\ h^{-1}}{4.2_0 \times 10^{-1} mol\ Pa^{-1} h^{-1}}$$

$$f_{Sys} = 3.8_1 \times 10^{-1} Pa$$

The concentration in the air and water are given by the product Zf:

$$C_W = Z_W f_{Sys} = 2.0_0 \times 10^{-2} mol\ Pa^{-1} m^{-3} \times 3.8_1 \times 10^{-1} Pa = 7.6_2 \times 10^{-3} mol\, m^{-3}$$

$$C_A = Z_A f_{Sys} = 4.0 \times 10^{-4} mol\ Pa^{-1} m^{-3} \times 3.8_1 \times 10^{-1} Pa = 1.5_2 \times 10^{-4} mol\, m^{-3}$$

The total molar outflow rates are given by GC:

$$r_W^{Outf} = G_W^{Outf} C_W$$

$$r_W^{Outf} = 1.0\ m^3 h^{-1} \times 7.6_2 \times 10^{-3} mol\, m^{-3}$$

$$r_W^{Outf} = 7.6_2 \times 10^{-3} mol\, h^{-1}$$

$$r_A^{Outf} = G_A^{Outf} C_A$$

$$r_A^{Outf} = 1.0 \times 10^3 m^3 h^{-1} \times 1.5_2 \times 10^{-4} mol\, m^{-3}$$

$$r_A^{Outf} = 1.5_2 \times 10^{-1} mol\, h^{-1}$$

As a check, we can sum the outflow rates to make sure it equals the inflow rate:

$$I = 1.6_0 \times 10^{-1}\, mol\ h^{-1}$$

$$r_{Total}^{Outf} = r_W^{Outf} + r_A^{Outf} \stackrel{?}{=} I$$

$$r_{Total}^{Outf} = 7.6_2 \times 10^{-3}\, mol\, h^{-1} + 1.5_2 \times 10^{-1}\, mol\, h^{-1} = 1.6_0 \times 10^{-1} mol\, h^{-1}$$

$$\therefore r_{Total}^{Outf} = r_{Total}^{Inf}$$

The total amount of the chemical in each compartment "i" is calculated as the product of $m_i = C_i V_i = V_i Z_i f_{sys}$:

$$m_W = V_W Z_W f_{Sys} = 5.0 \times 10^2 \times 2.0_0 \times 10^{-2} mol\ Pa^{-1} m^{-3} \times 3.8_1 \times 10^{-1}\, Pa = 3.8_1\, mol$$

$$m_A = V_A Z_A f_{Sys} = 5.0 \times 10^3 m^3 \times 4.0 \times 10^{-4} mol\ Pa^{-1} m^{-3} \times 3.8_1 \times 10^{-1}\, Pa = 7.6_2 \times 10^{-1}\, mol$$

The use of the D-value approach to flowing systems offers an alternative means to calculate rates of chemical flow directly from the fugacity in that compartment. For a given compartment at steady state,

$$r_i^{Adv} = r_i^{Inf} = r_i^{Outf} = G_i^{Adv} C_i = G_i^{Adv} Z_i f_i = D_i^{Adv} f_i$$

$$\boxed{r_i^{Adv} = D_i^{Adv} f_i}$$

Worked Example 3.4

Recalculate the rates of chemical outflow in the air and water from the previous example in terms of D-values.

We have the D-values already as well as the system fugacity, so this is a simple calculation:

$$r_W^{Outf} = D_W^{Adv} f_{Sys}$$

$$r_W^{Outf} = 2.0_0 \times 10^{-2} mol\ Pa^{-1} h^{-1} \times 3.8_1 \times 10^{-1}\, Pa = 7.6 \times 10^{-3}\, mol\ h^{-1}$$

$$r_A^{Outf} = D_A^{Adv} f_{Sys}$$

$$r_A^{Outf} = 4.0_0 \times 10^{-1} mol\ Pa^{-1} h^{-1} \times 3.8_1 \times 10^{-1}\, Pa = 1.5 \times 10^{-1}\, mol\ h^{-1}$$

3.2 BULK ADVECTION OF PARTICULATE MATTER

Advection into or out of an evaluative environment may also involve the transport of discrete particles containing a chemical suspended within a moving medium that itself may not contain the chemical, or may contain it at very different concentrations. Examples include the transport of aerosols in air and suspended particles and biota in water. The treatment of such transport is done in terms of the flow rate of the particulate G_P, as determined by product of the flow rate of the medium G_M in which it is contained and the volume fraction of the particulate matter in the medium v_M^{f-P}:

$$G_P = G_M v_M^{f-P}$$

Worked Example 3.5

Air containing aerosol particles (Q) is flowing into an air compartment at a rate of $5.0 \times 10^{10}\, m^3\, h^{-1}$. The volume fraction of aerosols in the air is $v_A^{f-Q} = 1.0 \times 10^{-12}$. What is the flow rate of aerosols into the air compartment? Use this flow rate and a chemical concentration in the aerosols of 10 $mol\, m^{-3}$ to determine the molar rate of chemical inflow into the air compartment due to the aerosols.

The aerosol flow rate is the product of the air flow rate and the volume fraction of aerosols:

$$G_Q = G_A v_A^{f-Q}$$

$$G_Q = 5.0 \times 10^{10} m^3 h^{-1} \times 1.0 \times 10^{-12} = 5.0_0 \times 10^{-2} m^3 h^{-1}$$

The molar inflow rate of chemical due to advected aerosol is given by *GC*:

$$r_Q^{Adv} = G_Q C_Q = 5.0_0 \times 10^{-2} m^3 h^{-1} \times 10.\ mol\ m^{-3} = 5.0_0 \times 10^{-1} mol\, h^{-1}$$

Worked Example 3.6

Assume a fugacity capacity of the aerosols of $Z_Q = 1.0 \times 10^6\ mol\ Pa^{-1}\ m^{-3}$ and that the system is at equilibrium. Use the *D*-value approach ($r = Df$) to determine the same molar flow rate for the aerosols as in Worked Example 3.5.

The molar flow rate is calculated from the advection *D*-value and the system fugacity, both of which we need to calculate.

The advection *D*-value is given by *GZ*:

$$D_Q^{Adv} = G_Q Z_Q$$

$$D_Q^{Adv} = 5.0_0 \times 10^{-2} m^3 h^{-1} \times 1.0 \times 10^6 mol\ Pa^{-1} m^{-3}$$

$$D_Q^{Adv} = 5.0_0 \times 10^4 mol\ Pa^{-1} h^{-1}$$

The system fugacity is given by *C/Z:*

$$f_{Sys} = \frac{C_Q}{Z_Q}$$

$$f_{Sys} = \frac{10.\ mol\ m^{-3}}{1.0 \times 10^6 mol\ Pa^{-1} m^{-3}} = 1.0_0 \times 10^{-5} Pa$$

The molar flow rate of chemical in the aerosols by the *D*-value approach is therefore:

$$r_Q^{Adv} = D_Q^{Adv} f_{Sys}$$

$$r_Q^{Adv} = 5.0_0 \times 10^4 mol\ Pa^{-1} h^{-1} \times 1.0_0 \times 10^{-5} Pa = 5.0 \times 10^{-1} mol\ h^{-1}$$

Worked Example 3.7

A lake is fed by a stream at a rate of $5.0 \times 10^2\ m^3\ h^{-1}$. The stream contains suspended solids with a volume fraction $v_W^{f-SS} = 3.0 \times 10^{-4}$. The suspended solids contain a chemical at a concentration of $5.0 \times 10^{-1}\ mol\ m^3$. What is the flow rate of suspended solids and of the chemical itself?

The suspended solids flow rate is simply the flow rate of the water, weighted by the volume fraction of suspended solids:

$$G_{SS} = G_W v_W^{f-SS}$$

$$G_{SS} = 5.0 \times 10^2 m^3 h^{-1} \times 3.0 \times 10^{-4} = 1.5_0 \times 10^{-1} m^3 h^{-1}$$

The molar inflow rate of chemical due to advected suspended solids is given by *GC*:

$$r_{SS}^{Adv} = G_{SS} C_{SS} = 1.5_0 \times 10^{-1} m^3 h^{-1} \times 5.0 \times 10^{-1}\ mol\ m^{-3} = 7.5 \times 10^{-2} mol\, h^{-1}$$

3.3 ADVECTION WITHIN THE EVALUATIVE ENVIRONMENT

Certain advective processes involve movement of particles containing a chemical of interest within a given evaluative environment. Many of these processes do not add or subtract chemical from the system but rather redistribute it amongst system compartments. Examples of such processes are rainfall, dry and wet particle advective transport from air to water or soil, suspended particle deposition from water to sediment, and soil matter and water runoff from soil to water. Figure 3.2 illustrates the three processes that apply to advective transfer from air to water.

D-values for such advective processes are defined in terms of an equivalent flow rate G ($m^3\ h^{-1}$). The equivalent flow rate is generally calculated as a product of a "velocity", U with units of length/time, and an area, A (m^2) over which the deposition takes place. Thus, for rainfall we have:

$$G_A^{Rain} = U^{Rain} \times A$$

If the deposition velocity refers to a medium in which the depositing material is entrained, an additional volume fraction needs to be incorporated. Thus, for dry deposition of particulate matter from air, where the deposition velocity refers to air containing the particles with a volume fraction v_A^{f-Dry}, we have:

$$G_A^{Dry} = U^{Dry} \times v_A^{f-Dry} \times A$$

Once the flow rate G is determined, the corresponding D-value is given by the product GZ, as with air and water advective transport introduced earlier.

Wet deposition is the process of aerosol scavenging by rain, which is treated separately from simple rainfall. For such a process, a scavenging ratio Q is employed, which corresponds to the volume of air that would need to be completely depleted of aerosols to account for the rate of aerosol deposition by rain scavenging, per unit volume of rain. The volume flow rate of aerosol scavenging from air by rain is therefore given by the volume flow rate of rain ($U \times A$) multiplied by the scavenging ratio, and by the volume fraction of aerosols in air, v_A^{f-Q}:

$$G_A^{Wet} = U^{Rain} \times A \times v_A^{f-Q} \times Q$$

Note that, since processes such as rainfall and particle deposition are not usually reported as a volume flow rate, it is necessary to employ various strategies and conversions to obtain flow rates in units of $m^3\ h^{-1}$.

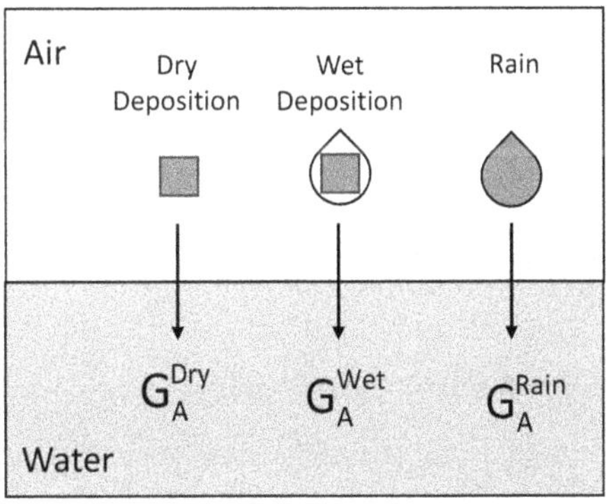

FIGURE 3.2 Illustration of the three main advective processes that result in air-to-water intercompartment transport within an evaluative environment.

For example, for rain reported in $mm\ yr^{-1}$, the rain flow rate in units of $m^3\ h^{-1}$ is given by:

$$G_A^{Rain}\left(m^3h^{-1}\right)=\frac{U^{Rain}\left(mm\,yr^{-1}\right)\times A\left(m^2\right)}{10^3\,mm\,m^{-1}\times 8760\,h\ yr^{-1}}$$

Worked Examples 3.8, 3.9, and 3.10 are based on data from Mackay et al. (1986), with the time unit converted from seconds to hours.

Worked Example 3.8

Rain falls in a region at a rate of 500 $mm\ yr^{-1}$. What is the rate of rainfall on a 1.00 m^2 area in units of $m^3\ h^{-1}$? What is the corresponding D-value for a chemical with a fugacity capacity in water of Z_W = 0.020 $mol\ Pa^{-1}\ m^{-3}$?

Using a model area of 1.00 m^2, the rate of rainfall is:

$$G_A^{Rain}\left(m^3h^{-1}\right)=\frac{U^{Rain}\left(mm\,yr^{-1}\right)\times A\left(m^2\right)}{10^3\,mm\,m^{-1}\times 8760\,h\ yr^{-1}}$$

$$G_A^{Rain}=\frac{500.\,mm\,yr^{-1}\times 1.00\,m^2}{10^3\,mm\,m^{-1}\times 8760\,h\ yr^{-1}}$$

$$G_A^{Rain}=5.71_0\times 10^{-5}m^3h^{-1}$$

The corresponding D-value is given by the product GZ:

$$D_A^{Rain}=G_A^{Rain}Z_W$$

$$D_A^{Rain}=5.71_0\times 10^{-5}m^3h^{-1}\times 0.020\ mol\ Pa^{-1}m^{-3}$$

$$D_A^{Rain}=1.1\times 10^{-6}mol\ Pa^{-1}h^{-1}$$

Worked Example 3.9

Particulate matter deposits from air at a velocity of 1.08 × 10[1] $m\ h^{-1}$. Calculate the dry deposition flow rate for a 1.00 m^2 area, assuming a volume fraction of particulate matter v_A^{f-Q} = 5.0 × 10^{-11}. What is the corresponding D-value for a chemical with a fugacity capacity in dry particulate matter Z_Q = 5.01 × 10^5 $mol\ Pa^{-1}\ m^{-3}$?

Using a model area of 1.00 m^2, the rate of dry deposition is:

$$G_A^{Dry}\left(m^3h^{-1}\right)=U^{Dry}\left(m\,h^{-1}\right)\times v_A^{f-Q}\times A\left(m^2\right)$$

$$G_A^{Dry}\left(m^3h^{-1}\right)=1.08\times 10^1 m\ h^{-1}\times 5.0\times 10^{-11}\times 1.00\ m^2$$

$$G_A^{Dry}\left(m^3h^{-1}\right)=5.4_0\times 10^{-10}m^3\ h^{-1}$$

The corresponding D-value is given by the product GZ:

$$D_A^{Dry}=G_A^{Dry}Z_Q$$

$$D_A^{Dry}=5.4_0\times 10^{-10}m^3\ h^{-1}\times 5.01\times 10^5 mol\ Pa^{-1}m^{-3}$$

$$D_A^{Dry}=2.7\times 10^{-4}mol\ Pa^{-1}h^{-1}$$

Worked Example 3.10

Aerosols are scavenged from air by rain with a scavenging ratio of $Q = 2.0 \times 10^5$. Assuming a volume fraction of aerosols in air of $v_A^{f-Q} = 5.0 \times 10^{-11}$ and a rain flow rate of $G_A^{Rain} = 5.71 \times 10^{-5} m^3 h^{-1}$, calculate the volume flow rate of aerosols by wet deposition (rain scavenging) on a 1.00 m^2 area. Use the value of $Z_Q = 5.01 \times 10^5$ $mol\ Pa^{-1} m^{-3}$ to calculate the corresponding D-value for wet deposition.

Since we already know the value of $U^{Rain} \times A = G_A^{Rain}$, the wet deposition rate is given by:

$$G_A^{Wet} = G_A^{Rain} \times v_A^{f-Q} \times Q$$

$$G_A^{Wet} = 5.71 \times 10^{-5} m^3 h^{-1} \times 5.0 \times 10^{-11} \times 2.0 \times 10^5$$

$$G_A^{Wet} = 5.7_1 \times 10^{-10} m^3 h^{-1}$$

The corresponding D-value is given by the product GZ:

$$D_A^{Wet} = G_A^{Wet} Z_Q$$

$$D_A^{Wet} = 5.7_1 \times 10^{-10} m^3 h^{-1} \times 5.01 \times 10^5 mol\ Pa^{-1} m^{-3}$$

$$D_A^{Wet} = 2.9 \times 10^{-4} mol\ Pa^{-1} h^{-1}$$

Sediment deposition, resuspension, and burial are all advective processes that may be active in a water column (See Figure 3.3), the last being a bulk advective transport process that removes sediment and hence some of the chemical from the evaluative environment as a whole.

The approach to estimating D-values for these processes is similar to that employed in air, with the key difference being that such processes are usually quantified in terms of a flux L, the mass of material undergoing the process per unit of area per unit of time (e.g., $kg\ m^{-2}\ h^{-1}$). To convert such a flux to a volume flow rate, the mass of material must be converted to its corresponding volume, and the area over which the process is occurring must be incorporated.

Thus, for deposition of suspended solids to the sediment, we have a volume flow rate defined as:

$$G_W^{Dep} = \frac{L_W^{Dep} \times A_{Sed-W}}{\rho_{SS}}$$

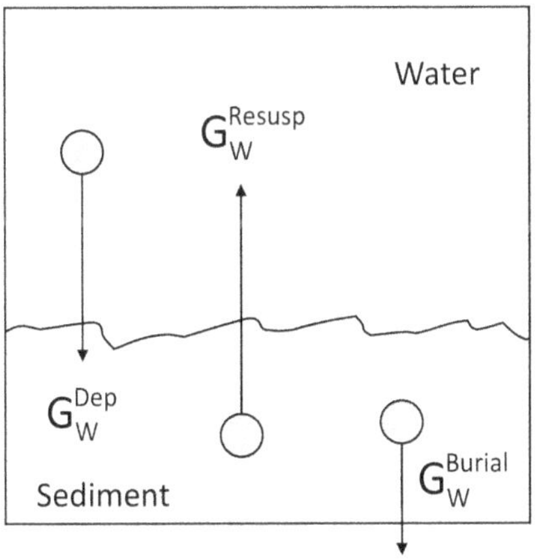

FIGURE 3.3 Schematic illustrating the three processes involved in particle transport of chemicals between water and sediment, as well as out of the system itself (burial).

For sediment resuspension and irreversible burial, the same equation form applies, but with the density of the deposited sediment instead of that of the suspended solids:

$$G_W^{Resusp} = \frac{L_W^{Resusp} \times A_{Sed-W}}{\rho_{Sed}}$$

$$G_W^{Burial} = \frac{L_W^{Burial} \times A_{Sed-W}}{\rho_{Sed}}$$

In the event that such fluxes are given in units other than $kg\ m^{-2}\ h^{-1}$, additional conversion factors must be included in the calculation. For example, for a suspended solids deposition flux in units of $g\ m^{-2}$ day^{-1}, the following conversions must be applied:

$$G_W^{Dep}\left(m^3 h^{-1}\right) = \frac{L_W^{Dep}\left(g\,m^{-2} day^{-1}\right) \times A_{Sed-W}\left(m^2\right)}{\rho_{SS}\left(kg\,m^{-3}\right) \times 24\,h\,day^{-1} \times 10^3\,g\,kg^{-1}}$$

Worked Example 3.11 is based on distribution of the insecticide dichlorodiphenyltrichloroethane (DDT) in the default environment of the Sediment model, available from the CEMC website (https://www.trentu.ca/cemc/resources-and-models).

Worked Example 3.11

A pond with an area of $1.0 \times 10^3\ m^2$ is contaminated with DDT. It undergoes suspended solids deposition with a flux of 3.0 $g\ m^{-2}\ day^{-1}$. The sediment resuspension flux is 1.0 $g\ m^{-2}\ day^{-1}$ and the sediment burial flux is 1.5 $g\ m^{-2}\ day^{-1}$. Determine the volume flow rates and corresponding D-values for sediment deposition, resuspension and burial, assuming a suspended solids density of 1497.4 $kg\ m^{-3}$ and a sediment density of 2204.74 $kg\ m^{-3}$. Use DDT fugacity capacities for suspended solids of $Z_{SS} = 8.44 \times 10^4\ mol\ Pa^{-1}\ m^{-3}$ and for sediment of $Z_{Sed} = 2.49 \times 10^4\ mol\ Pa^{-1}\ m^{-3}$.

The suspended solids deposition rate is:

$$G_W^{Dep} = \frac{L_W^{Dep}\left(g\,m^{-2} day^{-1}\right) \times A_{Sed-W}\left(m^2\right)}{\rho_{SS}\left(kg\,m^{-3}\right) \times 24\,h\,day^{-1} \times 10^3\,g\,kg^{-1}}$$

$$G_W^{Dep} = \frac{3.0\,g\,m^{-2} day^{-1} \times 1.0 \times 10^3 m^2}{1497.4\,kg\,m^{-3} \times 24\,h\,day^{-1} \times 10^3\,g\,kg^{-1}}$$

$$G_W^{Dep} = 8.3_5 \times 10^{-5} m^3 h^{-1}$$

The resuspension rate is:

$$G_W^{Resusp} = \frac{L_W^{Resusp}\left(g\,m^{-2} day^{-1}\right) \times A_{Sed-W}\left(m^2\right)}{\rho_{Sed}\left(kg\,m^{-3}\right) \times 24\,h\,day^{-1} \times 10^3\,g\,kg^{-1}}$$

$$G_W^{Resusp} = \frac{1.0\,g\,m^{-2} day^{-1} \times 1.0 \times 10^3 m^2}{2204.74\,kg\,m^{-3} \times 24\,h\,day^{-1} \times 10^3\,g\,kg^{-1}}$$

$$G_W^{Resusp} = 1.8_9 \times 10^{-5} m^3 h^{-1}$$

The burial rate is:

$$G_W^{Burial} = \frac{L_W^{Burial}\left(g\,m^{-2}day^{-1}\right)\times A_{Sed-W}\left(m^2\right)}{\rho_{Sed}\left(kg\,m^{-3}\right)\times 24\,h\,day^{-1}\times 10^3\,g\,kg^{-1}}$$

$$G_W^{Burial} = \frac{1.5\,g\,m^{-2}day^{-1}\times 1.0\times 10^3\,m^2}{2204.74\,kg\,m^{-3}\times 24\,h\,day^{-1}\times 10^3\,g\,kg^{-1}}$$

$$G_W^{Burial} = 2.8_3\times 10^{-5}\,m^3h^{-1}$$

The *D*-values for these three processes follow from the product of these volume flow rates and the fugacity capacity of the transported substance, either suspended solids (deposition) or sediment (resuspension and burial).

For deposition of suspended solids:

$$D_W^{Dep} = G_W^{Dep}\times Z_{SS}$$

$$D_W^{Dep} = 8.3_5\times 10^{-5}\,m^3h^{-1}\times 8.44\times 10^4\ mol\ Pa^{-1}m^{-3}$$

$$D_W^{Dep} = 7.0\ mol\ Pa^{-1}h^{-1}$$

For resuspension of sediment:

$$D_W^{Resusp} = G_W^{Resusp}\times Z_{Sed}$$

$$D_W^{Resusp} = 1.8_9\times 10^{-5}\,m^3h^{-1}\times 2.49\times 10^4\ mol\ Pa^{-1}m^{-3}$$

$$D_W^{Resusp} = 4.7\times 10^{-1}\ mol\ Pa^{-1}h^{-1}$$

For sediment burial:

$$D_W^{Burial} = G_W^{Burial}\times Z_{Sed}$$

$$D_W^{Burial} = 2.8_3\times 10^{-5}\,m^3h^{-1}\times 2.49\times 10^4\,mol\,Pa^{-1}m^{-3}$$

$$D_W^{Burial} = 7.1\times 10^{-1}\,mol\,Pa^{-1}h^{-1}$$

3.4 DEGRADATION PROCESSES

A given chemical may be transformed into one or more distinct products through chemical reactions. From the point of view of the chemical of interest, such degradation corresponds to a removal process in which the chemical "disappears" from the system. Notwithstanding the fact that the product of such degradation may be of equal or even greater concern compared with the original chemical of interest, from the point of view of modelling, the original chemical simply "disappears" from the system. The one exception occurs when an additional mechanism results in a "feedback" process that regenerates the initial chemical, but this is not normally the case.

Many degradation processes exhibit kinetics that are either first-order or pseudo-first-order, in nature. Here we will concern ourselves only with such processes, and for the present disregard higher-order processes which are less common, especially for environmental contaminants that are present in low concentrations.

For a first-order process with a rate constant k ($mol\,h^{-1}$) the rate of loss from a system or compartment is given by:

$$r = m\times k$$

Since the total moles of chemical in a system or compartment is given by the product of its volume multiplied by its concentration, we have:

$$\boxed{r_i^{Deg} = V_i C_i k_i^{Deg}}$$

Worked Example 3.12

DDT degrades in sediment by reaction with a first-order rate constant $k_{Sed}^{Deg} = 5.78 \times 10^{-7}\ h^{-1}$. Calculate the rate of degradation in a sediment body of $V_{Sed} = 30.0\ m^3$ and $C_{Sed} = 1.24 \times 10^{-2}\ mol\ m^{-3}$.

The rate of degradation follows directly from the product *VCk*:

$$r_{Sed}^{Deg} = V_{Sed} C_{Sed} k_{Sed}^{Deg}$$

$$r_{Sed}^{Deg} = 30.0\,m^3 \times 1.24 \times 10^{-2} mol\ m^{-3} \times 5.78 \times 10^{-7} h^{-1}$$

$$r_{Sed}^{Deg} = 2.1_5 \times 10^{-7} mol\,h^{-1}$$

Expressed in terms of fugacity, we have:

$$r_i^{Deg} = V_i \left(Z_i f_{Sys}\right) k_i^{Deg}$$

or

$$r_i^{Deg} = \left(V_i Z_i k_i^{Deg}\right) f_{Sys}$$

The product *VZk* has the same units as the advection *D*-value ($mol\ Pa^{-1}\ h^{-1}$) introduced above. We can therefore define:

$$\boxed{D_i^{Deg} = V_i Z_i k_i^{Deg}}$$

Expressing the term in parentheses as a *D*-value, we arrive at the same format for this rate as with advection processes for systems at equifugacity:

$$\boxed{r_i^{Deg} = D_i^{Deg} f_{Sys}}$$

The correspondence between the rate expression for advection and degradation when expressed in terms of fugacity is not a coincidence. Rather, it demonstrates one of the great strengths of using the fugacity approach compared with concentration-based theory, in that different transport rates can be directly compared "on equal footing", despite the fact that the physical processes to which they correspond are qualitatively different. We shall see that this advantage is not limited to just these two processes. By employing fugacity, all transport processes will be on such equal footing, with the same units, and will be directly comparable. This advantage gives the user the ability to directly compare disparate physical processes in terms of their relative contribution to the overall movement and distribution of a chemical within a model environment.

Worked Example 3.13

Use a fugacity capacity of sediment for DDT of $Z_{Sed} = 4976.\ mol\ Pa^{-1}\ m^{-3}$ to calculate the degradation *D*-value for DDT in sediment ($V_{Sed} = 30.0\ m^3$) then use a fugacity of $f_{Sed} = 2.49 \times 10^{-6}\ Pa$ to recalculate the degradation rate from Worked Example 3.12. (Recall for DDT in sediment, we used $k_{Sed}^{Deg} = 5.78 \times 10^{-7}\ h^{-1}$.)

The D-value for degradation follows directly from the product VZk:

$$D_{Sed}^{Deg} = V_{Sed} \times Z_{Sed} \times k_{Sed}^{Deg}$$
$$D_{Sed}^{Deg} = 30.0\,m^3 \times 4976.\,mol\ Pa^{-1}m^{-3} \times 5.78 \times 10^{-7}\,h^{-1}$$
$$D_{Sed}^{Deg} = 8.62_8 \times 10^{-2}\,mol\ Pa^{-1}h^{-1}$$

The rate of degradation now follows directly from the product Df:

$$r_{Sed}^{Deg} = D_{Sed}^{Deg} \times f_{Sed}$$
$$r_{Sed}^{Deg} = 8.62_8 \times 10^{-2}\,mol\ Pa^{-1}h^{-1} \times 2.49 \times 10^{-6}\,Pa$$
$$r_{Sed}^{Deg} = 2.1 \times 10^{-7}\,mol\ h^{-1}$$

When two or more degradation processes are present in an equilibrating system or compartment, each process is assigned its own D-value, and the total loss rate due to all processes is the sum of these D-values for degradation multiplied by the prevailing fugacity. This follows since all parallel first-order rate constants are additive. For n compartments at equilibrium, we have:

$$r^{Deg} = \sum_{i=1}^{n} V_i C_i k_i^{Deg} = \sum_{i=1}^{n} V_i Z_i f_{Sys} k_i^{Deg} = f_{Sys}\left(\sum_{i=1}^{n} V_i Z_i k_i^{Deg}\right)$$

$$\boxed{r^{Deg} = f_{Sys} \sum_{i=1}^{n} D_i^{Deg}}$$

For a system in which only degradation occurs, but no advection, we then have an expression for the system fugacity:

$$\boxed{f_{Sys} = \frac{r^{Deg}}{\sum_{i=1}^{n} D_i^{Deg}}}$$

The following is a Worked Example based on Test Chemical 1 in the EQC standard environment, as implemented in the EQC model available from the CEMC website (https://www.trentu.ca/cemc/resources-and-models). Figure 3.4 illustrates the various degradation processes involved.

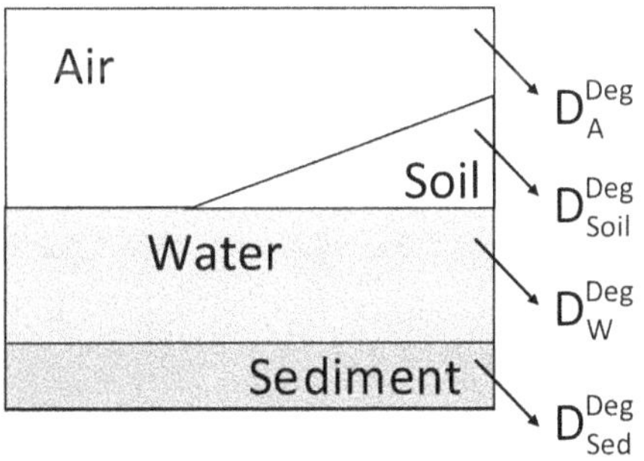

FIGURE 3.4 Illustration of four primary degradation processes for a four-compartment evaluative environment, and their associated D-value symbols.

Worked Example 3.14

A chemical has the following degradation rate constants in the EQC standard environment, for which the compartment volumes and fugacity capacities are also given:

Compartment	V_i (m^3)	Z_i ($mol\ Pa^{-1}\ m^{-3}$)	k_i^{Deg} (h^{-1})
Air	1.00×10^{14}	4.03×10^{-4}	4.08×10^{-2}
Water	2.00×10^{11}	6.00	1.26×10^{-2}
Soil	9.00×10^{9}	37.3	4.08×10^{-3}
Sediment	1.00×10^{8}	74.4	4.08×10^{-3}

The four compartments have individual D-values as illustrated in Figure 3.4. Assuming all compartments are at equilibrium and the equilibrium fugacity is $2.02 \times 10^{-7}\ Pa$, calculate the overall rate of loss of chemical by degradation.

The individual D-values are calculated as VZk, and the total degradation rate can be calculated by the product of the sum of D-values for individual compartment degradations multiplied by the system fugacity ΣDf. Thus,

The D-value for degradation in air is:

$$D_A^{Deg} = V_A Z_A k_A^{Deg}$$

$$D_A^{Deg} = 1.00 \times 10^{14} m^3 \times 4.03 \times 10^{-4} mol\ Pa^{-1} m^{-3} \times 4.08 \times 10^{-2} h^{-1}$$

$$D_A^{Deg} = 1.64_4 \times 10^9 mol\ Pa^{-1} h^{-1}$$

The analogous calculations for the remaining three degradation D-values are similar:

$$D_W^{Deg} = V_W \times Z_W \times k_W^{Deg} = 2.00 \times 10^{11} m^3 \times 6.00\ mol\ Pa^{-1} m^{-3} \times 1.26 \times 10^{-2} h^{-1} = 1.51_2 \times 10^{10} mol\ Pa^{-1} h^{-1}$$

$$D_{Soil}^{Deg} = V_{Soil} \times Z_{Soil} \times k_{Soil}^{Deg} = 9.00 \times 10^9 m^3 \times 37.3\ mol\ Pa^{-1} m^{-3} \times 4.08 \times 10^{-3} h^{-1} = 1.37_0 \times 10^9 mol\ Pa^{-1} h^{-1}$$

$$D_{Sed}^{Deg} = V_{Sed} \times Z_{Sed} \times k_{Sed}^{Deg} = 1.00 \times 10^8 m^3 \times 74.4\ mol\ Pa^{-1} m^{-3} \times 4.08 \times 10^{-3} h^{-1} = 3.03_6 \times 10^7 mol\ Pa^{-1} h^{-1}$$

The sum of degradation D-values for the system is:

$$\sum_{i=1}^{n} D_i^{Deg} = D_A^{Deg} + D_W^{Deg} + D_{Soil}^{Deg} + D_{Sed}^{Deg}$$

$$\sum_{i=1}^{n} D_i^{Deg} = \left(1.64_4 \times 10^9 + 1.51_2 \times 10^{10} + 1.37_0 \times 10^9 + 3.03_6 \times 10^7\right) mol\ Pa^{-1} h^{-1}$$

$$\sum_{i=1}^{n} D_i^{Deg} = 1.81_6 \times 10^{10} mol\ Pa^{-1} h^{-1}$$

Therefore, the overall or total rate for loss due to degradation in the system is given by:

$$r^{Deg} = f_{Sys} \sum_{i=1}^{n} D_i^{Deg}$$

$$r^{Deg} = 2.02 \times 10^{-7} Pa \times 1.81_6 \times 10^{10} mol\ Pa^{-1} h^{-1}$$

$$r^{Deg} = 3.67 \times 10^3 mol\ h^{-1}$$

The loss rates due to degradation in each of the compartments individually are given by the product *fD*:

$$r_A^{Deg} = f_{Sys} D_A^{Deg} = 2.02 \times 10^{-7}\,Pa \times 1.64_4 \times 10^9\,mol\,Pa^{-1}h^{-1} = 332._1\,mol\,h^{-1}$$

$$r_W^{Deg} = f_{Sys} D_W^{Deg} = 2.02 \times 10^{-7}\,Pa \times 1.51_2 \times 10^{10}\,mol\,Pa^{-1}h^{-1} = 3.05_4 \times 10^3\,mol\,h^{-1}$$

$$r_{Soil}^{Deg} = f_{Sys} D_{Soil}^{Deg} = 2.02 \times 10^{-7}\,Pa \times 1.37_0 \times 10^9\,mol\,Pa^{-1}h^{-1} = 276._7\,mol\,h^{-1}$$

$$r_{Sed}^{Deg} = f_{Sys} D_{Sed}^{Deg} = 2.02 \times 10^{-7}\,Pa \times 3.03_6 \times 10^7\,mol\,Pa^{-1}h^{-1} = 6.13_2\,mol\,h^{-1}$$

As a check, these should add up to the total degradation rate, as they must:

$$r^{Deg} = \sum_{i=1}^{n} r_i^{Deg} = r_A^{Deg} + r_W^{Deg} + r_{Soil}^{Deg} + r_{Sed}^{Deg}$$

$$r^{Deg} = 332._1 + 3.05_4 \times 10^3 + 276._7 + 6.13_2 = 3.67 \times 10^3\,mol\,h^{-1}$$

In this case, the result illuminates the relative importance of the degradation in water compared to the other compartments, driven by a combination of the relatively high volume of water and the relatively high degradation rate of the chemical in question in water.

3.5 COMBINING ADVECTION AND DEGRADATION PROCESSES

Advection and degradation processes are similar from the point of view that their rates are defined as products of *D*-values and fugacity. As a result, systems in which both advection and degradation occur simultaneously can be easily treated with little additional theory. The principal relation recognizes that the total rate of loss for a chemical in a compartment or system due to these non-diffusive processes is simply the sum of the total advection and degradation rates.

It follows directly that the *D*-value associated with the sum of both these non-diffusive loss processes is simply the sum of the *D*-values for the individual processes:

$$r^L = \sum_{i=1}^{n} r_i^{Adv} + \sum_{i=1}^{n} r_i^{Deg}$$

$$D^L f_{Sys} = \sum_{i=1}^{n} D_i^{Adv} f_{Sys} + \sum_{i=1}^{n} D_i^{Deg} f_{Sys}$$

$$D^L = \sum_{i=1}^{n} D_i^{Adv} + \sum_{i=1}^{n} D_i^{Deg}$$

Note that the rates of loss due to various advection and degradation reactions may all be compared through the relative magnitude of the associated *D*-values for these processes. The dominant loss process(es) will be the one(s) with the largest *D*-value(s).

Worked Example 3.15

An air body of volume $1.0 \times 10^{14}\,m^3$ has advective flow input at a rate of $1.0 \times 10^{12}\,m^3\,h^{-1}$. The air contains a chemical whose fugacity capacity in air is $Z_A = 4.03 \times 10^{-4}\,mol\,Pa^{-1}\,m^{-3}$. The fugacity in the air is $2.02 \times 10^{-7}\,Pa$. The chemical undergoes first-order degradation with a rate constant of $4.08 \times 10^{-2}\,h^{-1}$. Calculate the rates of advection and degradation, then combine the *D*-values for these two processes and calculate the total rate of loss for these two processes combined.

To determine the rates of chemical advection and degradation, we need to know the *D*-values for advection and degradation.

The *D*-value for advection is given by *GZ*:

$$D_A^{Adv} = G_A^{Inf} Z_A$$

$$D_A^{Adv} = 1.0 \times 10^{12} m^3 h^{-1} \times 4.03 \times 10^{-4} mol\ Pa^{-1} m^{-3}$$

$$D_A^{Adv} = 4.0_3 \times 10^8 mol\ Pa^{-1} h^{-1}$$

The *D*-value for degradation is given by *VZk:*

$$D_A^{Deg} = V_A Z_A k_A^{Deg}$$

$$D_A^{Deg} = 1.0 \times 10^{14} m^3 \times 4.03 \times 10^{-4}\ mol\ Pa^{-1} m^{-3} \times 4.08 \times 10^{-2} h^{-1}$$

$$D_A^{Deg} = 1.6_4 \times 10^9\ mol\ Pa^{-1} h^{-1}$$

The rate of advective loss is given by $D^{adv}f$:

$$r_A^{Adv} = D_A^{Adv} \times f_A$$

$$r_A^{Adv} = 4.0_3 \times 10^8 mol\ Pa^{-1} h^{-1} \times 2.02 \times 10^{-7} Pa$$

$$r_A^{Adv} = 8.1_4 \times 10^1 mol\ h^{-1}$$

The rate of degradation loss is given by $D^{deg}f$:

$$r_A^{Deg} = D_A^{Deg} \times f_A$$

$$r_A^{Deg} = 1.6_4 \times 10^9\ mol\ Pa^{-1} h^{-1} \times 2.02 \times 10^{-7} Pa$$

$$r_A^{Deg} = 3.3_2 \times 10^2 mol\ h^{-1}$$

Finally, the total rate of loss based on the product of the sum of *D*-values and the fugacity:

$$D^L = D_A^{Adv} + D_A^{Deg} = 4.0_3 \times 10^8 mol\ Pa^{-1} h^{-1} + 1.6_4 \times 10^9\ mol\ Pa^{-1} h^{-1} = 2.0_5 \times 10^9\ mol\ Pa^{-1} h^{-1}$$

$$r^L = D^L f_A = 2.0_5 \times 10^9\ mol\ Pa^{-1} h^{-1} \times 2.02 \times 10^{-7} Pa = 4.1 \times 10^2\ mol\ h^{-1}$$

Note that the sum of the individual loss process rates is also 4.1×10^2 $mol\ h^{-1}$, as required. Also, the rate of loss due to degradation is about 4 times as great as that for advective loss, in accordance with the ratio of *D*-values for these processes, which is 4.1.

The fugacity for an open system at steady state undergoing both advection and degradation losses is given by rearrangement of the following mass flow balance:

$$I = \sum_{i=1}^{n} G_i^{Outf} C_i + \sum_{i=1}^{n} V_i C_i k_i^{Deg}$$

$$I = \sum_{i=1}^{n} G_i^{Outf} Z_i f_{Sys} + \sum_{i=1}^{n} V_i Z_i f_{Sys} k_i^{Deg}$$

$$f_{Sys} = \frac{I}{\sum_{i=1}^{n} G_i^{Outf} Z_i + \sum_{i=1}^{n} V_i Z_i k_i^{Deg}}$$

Given that both denominator terms are sums of D-values, we can compactly write:

$$f_{Sys} = \frac{I}{\sum_{i=1}^{n} D_i^{Adv} + \sum_{i=1}^{n} D_i^{Deg}}$$

$$\boxed{f_{Sys} = \frac{I}{\sum_{i=1}^{n} D_i^{L}}}$$

3.6 RESIDENCE TIMES FOR ADVECTIVE AND DEGRADATION LOSS PROCESSES

A key computable characteristic of steady-state systems is the residence time of a chemical in a particular compartment, and in the system as a whole. For the system as a whole, this is the average time a chemical spends in the system, and it is given by:

$$\boxed{\tau = \frac{m}{I}}$$

where m is total moles of chemical in the system, and, as before, I is the total rate of chemical input into the system. For systems involving only advective transport, $I = r^{Adv}$ and:

$$\boxed{\tau^{Adv} = \frac{m}{r^{Adv}}}$$

The advection residence times for individual compartments is given by an analogous expression involving the amount of chemical in the compartment and the rate of chemical flow into, or out of, that compartment:

$$\tau_i^{Adv} = \frac{m_i}{I_i} = \frac{m_i}{r_i^{Adv}}$$

The advection residence time for a given compartment is also related to its advective flow rate and total volume. Figure 3.5 illustrates this relationship graphically, and may be derived as follows:

$$\tau_i^{Adv} = \frac{m_i}{r_i^{Adv}} = \frac{C_i V_i}{G_i^{Adv} C_i} = \frac{V_i}{G_i^{Adv}}$$

$$\boxed{\tau_i^{Adv} = \frac{V_i}{G_i^{Adv}}}$$

Thus, the advective flow rate and advection residence time for a compartment are closely related via the compartment volume, and each can be easily determined from the other. As a result, it is not uncommon to quote advection residence times instead of advection flow rates, since the former is more intuitively useful in terms of "eyeballing" the importance of advection on the movement of the chemical through the compartment and system.

Figure 3.5 is a graphical illustration of the relationship between compartment volume, advective flow rate, and advective residence time, in which one "unit" of volume is added per hour to a volume of 12 such units, resulting in a residence time of 12 hours.

$$\tau_A^{Adv} = \frac{\text{Total volume}}{\text{Advective flow}} = \frac{12\,\text{units}}{1.0\,\text{units}\,\text{h}^{-1}} = 12\,\text{h}$$

FIGURE 3.5 Illustration of the relationship between compartment volume, advective flow rate, and advective residence time. Here one "unit" per hour of volume per enters and leaves a compartment with a volume of 12 such units, resulting in a residence time of 12 hours.

Advection residence time for the system as a whole may also be expressed in terms of D-values, according to the following derivation:

$$\tau^{Adv} = \frac{m}{r_{Total}^{Adv}} = \frac{\sum_{i=1}^{n} C_i V_i}{\sum_{i=1}^{n} G_i^{Adv} C_i} = \frac{\sum_{i=1}^{n} V_i Z_i f_i}{\sum_{i=1}^{n} G_i^{Adv} Z_i f_i} = \frac{\sum_{i=1}^{n} V_i Z_i}{\sum_{i=1}^{n} D_i^{Adv}}$$

$$\boxed{\tau^{Adv} = \frac{\sum_{i=1}^{n} V_i Z_i}{\sum_{i=1}^{n} D_i^{Adv}}}$$

Accordingly, the advection residence time for a given compartment expressed in terms of the D-value for advective flow is given by:

$$\boxed{\tau_i^{Adv} = \frac{m_i}{r_i^{Adv}} = \frac{V_i Z_i}{D_i^{Adv}}}$$

Worked Example 3.16

Using data and results from Worked Example 3.3, calculate the total amount of chemical in that system. Then, determine the advection residence time of the chemical in the system as a whole, as well as separately for the air and water.

The total amount of chemical in the system is:

$$m = m_W + m_A$$

$$m = 3.8_1\,mol + 7.6_2 \times 10^{-1}\,mol$$

$$m = 4.5_7\,mol$$

Therefore, given that the total molar inflow was found to be $I = 1.6_0 \times 10^{-1} mol\ h^{-1}$, the advection residence time for the system as a whole is:

$$\tau = \frac{m}{I} = \frac{4.5_7\ mol}{1.6_0 \times 10^{-1} mol\ h^{-1}} = 28._6\ h$$

The advection residence time for the individual compartments is given by:

$$\tau_W = \frac{m_W}{r_W^{Adv}} = \frac{3.8_1\ mol}{7.6_2 \times 10^{-3}\ mol\ h^{-1}} = 5.0_0 \times 10^2\ h$$

$$\tau_A = \frac{m_A}{r_A^{Adv}} = \frac{7.6_2 \times 10^{-1}\ mol}{1.5_2 \times 10^{-1}\ mol\ h^{-1}} = 5.0_0\ h$$

Using the D-value approach, we can arrive at the same result:

$$\tau = \frac{\sum_{i=1}^{n} V_i Z_i}{\sum_{i=1}^{n} D_i^{Adv}} = \frac{V_W Z_W + V_A Z_A}{D_W^{Adv} + D_A^{Adv}}$$

$$\tau = \frac{5.00 \times 10^2 m^3 \times 2.0 \times 10^{-2} mol\ Pa^{-1} m^{-3} + 5.0 \times 10^3 m^3 \times 4.0 \times 10^{-4} mol\ Pa^{-1} m^{-3}}{2.0_0 \times 10^{-2} mol\ Pa^{-1} h^{-1} + 4.0_0 \times 10^{-1} mol\ Pa^{-1} h^{-1}} = 28._6\ h$$

$$\tau_W = \frac{V_W Z_W}{D_W^{Adv}} = \frac{5.00 \times 10^2 m^3 \times 2.0 \times 10^{-2} mol\ Pa^{-1} m^{-3}}{2.0_0 \times 10^{-2} mol\ Pa^{-1} h^{-1}} = 5.0_0 \times 10^2\ h$$

$$\tau_A = \frac{V_A Z_A}{D_A^{Adv}} = \frac{5.00 \times 10^3 m^3 \times 4.0 \times 10^{-4} mol\ Pa^{-1} m^{-3}}{4.0_0 \times 10^{-1} mol\ Pa^{-1} h^{-1}} = 5.0_0\ h$$

Residence times may also be evaluated for degradation processes, in which case they refer to the average length of time a chemical exists in a system when subject to one or more first-order degradation removal processes. In the absence of advective flow, the rate of total molar inflow I is equal to the direct emission rate E. For a steady-state system, the total rate of loss due to degradation r^{Deg} must also equal the rate of inflow I, such that $I = E = r^{Deg}$.

In this case (degradation only), the system residence time for n compartments at steady state is given by:

$$\tau^{Deg} = \frac{m}{I} = \frac{m}{E} = \frac{m}{r^{Deg}} = \frac{\sum_{i=1}^{n} m_i}{\sum_{i=1}^{n} r_i^{Deg}}$$

For individual compartments, it is possible to define a theoretical residence time that would apply if the degradation occurred in that compartment as an isolated unit:

$$\tau_i^{Deg} = \frac{m_i}{r_i^{Deg}}$$

It is also possible to calculate the residence time for systems that involve losses by both advection and degradation operating in parallel. When advection and degradation are present in an open system at steady state, the overall rate for loss of chemical from the system as a whole r^L is given as simple sums of the individual compartment rates:

$$r^L = r^{Adv} + r^{Deg} = \sum_{i=1}^{n} r_i^{Adv} + \sum_{i=1}^{n} r_i^{Deg}$$

Given the reciprocal nature of residence time and rate of loss, we can rewrite this expression as:

$$\frac{m}{\tau^L} = \frac{m}{\tau^{Adv}} + \frac{m}{\tau^{Deg}} = \sum_{i=1}^{n} \frac{m_i}{\tau_i^{Adv}} + \sum_{i=1}^{n} \frac{m_i}{\tau_i^{Deg}}$$

Cancelling the total moles of chemical m (or the sum of moles in all compartments), we arrive at:

$$\frac{1}{\tau^L} = \frac{1}{\tau^{Adv}} + \frac{1}{\tau^{Deg}} = \sum_{i=1}^{n} \frac{1}{\tau_i^{Adv}} + \sum_{i=1}^{n} \frac{1}{\tau_i^{Deg}}$$

It is clear that residence times add "reciprocally" to give the reciprocal of the total residence time for the system. As a result, contributions from compartments with relatively short residence time will dominate in the sum, and compartments with relatively long residence times will not affect the system residence time significantly.

Worked Example 3.17

The following data corresponds to the EQC standard Level II environment. Use this data to determine the individual rates of loss from air, water, soil, and sediment, as well as the theoretical residence time for each compartment, the residence time for total advection and total degradation, and finally the residence time for all loss processes in the system as a whole.

Compartment	V_i (m^3)	Z_i (mol Pa^{-1} m^{-3})	G_i^{Adv} (m^3 h^{-1})	k_i^{Deg} (h^{-1})	C_i (mol m^{-3})
Air	1.00×10^{14}	4.03×10^{-4}	1.00×10^{12}	4.08×10^{-2}	8.155×10^{-11}
Water	2.00×10^{11}	6.00	2.00×10^{8}	1.26×10^{-2}	1.214×10^{-6}
Soil	9.00×10^{9}	37.3	–	4.08×10^{-3}	7.548×10^{-6}
Sediment	1.00×10^{8}	74.4	2.00×10^{3}	4.08×10^{-3}	1.505×10^{-5}

The rates of loss due to advection from each compartment are given by $r^{Adv} = G^{Adv}C$:

$$r_A^{Adv} = G_A^{Adv} C_A = 1.00 \times 10^{12} m^3 h^{-1} \times 8.155 \times 10^{-11} mol\, m^{-3} = 81.5_5\, mol\, h^{-1}$$
$$r_W^{Adv} = G_W^{Adv} C_W = 2.00 \times 10^{8} m^3 h^{-1} \times 1.214 \times 10^{-6} mol\, m^{-3} = 242._8 mol\, h^{-1}$$
$$r_{Sed}^{Adv} = G_{Sed}^{Adv} C_{Sed} = 2.00 \times 10^{3} m^3 h^{-1} \times 1.505 \times 10^{-5} mol\, m^{-3} = 3.01_0 \times 10^{-2} mol\, h^{-1}$$

The rates of loss due to degradation from each compartment are given by $r^{Deg} = VCk^{Deg}$:

$$r_A^{Deg} = V_A C_A k_A^{Deg} = 1.00 \times 10^{14} m^3 \times 8.155 \times 10^{-11} mol\, m^{-3} \times 4.08 \times 10^{-2} h^{-1} = 332._7\, mol\, h^{-1}$$
$$r_W^{Deg} = V_W C_W k_W^{Deg} = 2.00 \times 10^{11} m^3 \times 1.214 \times 10^{-6} mol\, m^{-3} \times 1.26 \times 10^{-2} h^{-1} = 3.05_9 \times 10^{3} mol\, h^{-1}$$
$$r_{Soil}^{Deg} = V_{Soil} C_{Soil} k_{Soil}^{Deg} = 9.00 \times 10^{9} m^3 \times 7.548 \times 10^{-6} mol\, m^{-3} \times 4.08 \times 10^{-3} h^{-1} = 277._2\, mol\, h^{-1}$$
$$r_{Sed}^{Deg} = V_{Sed} C_{Sed} k_{Sed}^{Deg} = 1.00 \times 10^{8} m^3 \times 1.505 \times 10^{-5} mol\, m^{-3} \times 4.08 \times 10^{-3} h^{-1} = 6.14_0\, mol\, h^{-1}$$

The residence times for advection and degradation are given by the amount of matter in each compartment divided by the appropriate rate of loss $\tau = m/r$. To use this, we need to know the amounts in each compartment, m_i:

$$m_A = V_A C_A = 1.00 \times 10^{14} m^3 \times 8.155 \times 10^{-11} mol\, m^{-3} = 8.15_5 \times 10^3 mol$$

$$m_W = V_W C_W = 2.00 \times 10^{11} m^3 \times 1.214 \times 10^{-6} mol\, m^{-3} = 2.42_8 \times 10^5 mol$$

$$m_{Soil} = V_{Soil} C_{Soil} = 9.00 \times 10^9 m^3 \times 7.548 \times 10^{-6} mol\, m^{-3} = 6.79_3 \times 10^4 mol$$

$$m_{Sed} = V_{Sed} C_{Sed} = 1.00 \times 10^8 m^3 \times 1.505 \times 10^{-5} mol\, m^{-3} = 1.50_5 \times 10^3 mol$$

$$\tau_A^{Adv} = \frac{m_A}{r_A^{Adv}} = \frac{8.15_5 \times 10^3 mol}{81.5_5\, mol\, h^{-1}} = 100._0 h$$

$$\tau_W^{Adv} = \frac{m_W}{r_W^{Adv}} = \frac{2.42_8 \times 10^5 mol}{242._8 mol\, h^{-1}} = 1.00_0 \times 10^3 h$$

$$\tau_{Sed}^{Adv} = \frac{m_{Sed}}{r_{Sed}^{Adv}} = \frac{1.50_5 \times 10^3 mol}{3.01_0 \times 10^{-2} mol\, h^{-1}} = 5.00_0 \times 10^4 h$$

$$\tau_A^{Deg} = \frac{m_A}{r_A^{Deg}} = \frac{8.15_5 \times 10^3 mol}{332._7\, mol\, h^{-1}} = 24.5_1 h$$

$$\tau_W^{Deg} = \frac{m_W}{r_W^{Deg}} = \frac{2.42_8 \times 10^5 mol}{3.05_9 \times 10^3 mol\, h^{-1}} = 79.3_7 h$$

$$\tau_{Soil}^{Deg} = \frac{m_{Soil}}{r_{Soil}^{Deg}} = \frac{6.79_3 \times 10^4 mol}{277._2\, mol\, h^{-1}} = 245._1 h$$

$$\tau_{Sed}^{Deg} = \frac{m_{Sed}}{r_{Sed}^{Deg}} = \frac{1.50_5 \times 10^3 mol}{6.14_0\, mol\, h^{-1}} = 245._1 h$$

The total residence time for advection and degradation in the whole system can be obtained from the sum of all rates for either type of loss and the relation $\tau = m/r$.

The total residence time for advection is:

$$\tau^{Adv} = \frac{m}{r^{Adv}} = \frac{\sum_{i=1}^{n} m_i}{\sum_{i=1}^{n} r_i^{Adv}}$$

$$\tau^{Adv} = \frac{3.20_4 \times 10^5 mol}{3.24_4 \times 10^2 mol\, h^{-1}} = 987._7 h$$

The total residence time for degradation is:

$$\tau^{Deg} = \frac{m}{r^{Deg}} = \frac{\sum_{i=1}^{n} m_i}{\sum_{i=1}^{n} r_i^{Deg}}$$

$$\tau^{Deg} = \frac{3.20_4 \times 10^5 mol}{3.67_5 \times 10^3 mol\, h^{-1}} = 87.1_7 h$$

The total residence time for the system associated with both advective and degradation loss can be calculated explicitly...

$$\tau^L = \frac{m}{r^L} = \frac{\sum_{i=1}^{n} m_i}{\sum_{i=1}^{n} r_i^{Adv} + \sum_{i=1}^{n} r_i^{Deg}}$$

$$\tau^L = \frac{3.20_4 \times 10^5 mol}{3.24_4 \times 10^2\, mol\, h^{-1} + 3.67_5 \times 10^3\, mol\, h^{-1}} = 80.1_0\, h$$

... or by way of adding the individual advection and degradation residence times reciprocally:

$$\frac{1}{\tau^L} = \frac{1}{\tau^{Adv}} + \frac{1}{\tau^{Deg}} = \frac{1}{987._7\, h} + \frac{1}{87.1_7\, h} = \frac{1}{80.1_0\, h}$$

Worked Example 3.18

Use D-values to recalculate the residence times determined in Worked Example 3.17 for advective and degradation losses from each of the compartments, for total advection and total degradation in the system, and finally the residence time for total losses for the system as a whole. Use the fugacity capacities from Worked Example 3.17.

In all cases, the residence time for a given volume subject to a single loss process is expressed in terms of a D-value for that loss as:

$$\tau = \frac{VZ}{D}$$

For advection and degradation occurring together within a given compartment, the residence time is given by:

$$\tau_i = \frac{V_i Z_i}{D_i^{Adv} + D_i^{Deg}}$$

The residence time for the system as a whole is given by:

$$\tau = \frac{\sum_{i=1}^{n} V_i Z_i}{\sum_{i=1}^{n} D_i^{Adv} + \sum_{i=1}^{n} D_i^{Deg}}$$

To proceed, we need to calculate the D-values for the various loss processes.

For advection, we use the relation $D = GZ$:

$$D_A^{Adv} = G_A^{Adv} Z_A = 1.00 \times 10^{12} m^3\, h^{-1} \times 4.03 \times 10^{-4}\, mol\, Pa^{-1}\, m^{-3} = 4.03_0 \times 10^8 mol\, Pa^{-1}\, h^{-1}$$

$$D_W^{Adv} = G_W^{Adv} Z_W = 2.00 \times 10^8 m^3\, h^{-1} \times 6.00\, mol\, Pa^{-1}\, m^{-3} = 1.20_0 \times 10^9 mol\, Pa^{-1}\, h^{-1}$$

$$D_{Sed}^{Adv} = G_{Sed}^{Adv} Z_{Sed} = 2.00 \times 10^3 m^3\, h^{-1} \times 74.4\, mol\, Pa^{-1}\, m^{-3} = 1.48_8 \times 10^5 mol\, Pa^{-1}\, h^{-1}$$

For degradation, we use the relation $D = VZk$:

$$D_A^{Deg} = V_A Z_A k_A^{Deg} = 1.00\times10^{14} m^3 \times 4.03\times10^{-4}\, mol\, Pa^{-1} m^{-3} \times 4.08\times10^{-2} h^{-1} = 1.64_4\times10^9\, mol\, Pa^{-1} h^{-1}$$

$$D_W^{Deg} = V_W Z_W k_W^{Deg} = 2.00\times10^{11} m^3 \times 6.00\, mol\, Pa^{-1} m^{-3} \times 1.26\times10^{-2} h^{-1} = 1.51_2\times10^{10}\, mol\, Pa^{-1} h^{-1}$$

$$D_{Soil}^{Deg} = V_{Soil} Z_{Soil} k_{Soil}^{Deg} = 9.00\times10^9 m^3 \times 37.3\, mol\, Pa^{-1} m^{-3} \times 4.08\times10^{-3} h^{-1} = 1.37_0\times10^9\, mol\, Pa^{-1} h^{-1}$$

$$D_{Sed}^{Deg} = V_{Sed} Z_{Sed} k_{Sed}^{Deg} = 1.00\times10^8 m^3 \times 74.4\, mol\, Pa^{-1} m^{-3} \times 4.08\times10^{-3} h^{-1} = 3.03_6\times10^7\, mol\, Pa^{-1} h^{-1}$$

The sum of all advective D-values is:

$$\sum_{i=1}^{n} D_i^{Adv} = D_A^{Adv} + D_W^{Adv} + D_{Sed}^{Adv}$$
$$= \left(4.03_0\times10^8 + 1.20_0\times10^9 + 1.48_8\times10^5\right) mol\, Pa^{-1} h^{-1}$$
$$= 1.60_3\times10^9\, mol\, Pa^{-1} h^{-1}$$

The sum of all degradation D-values is:

$$\sum_{i=1}^{n} D_i^{Deg} = D_A^{Deg} + D_W^{Deg} + D_{Soil}^{Deg} + D_{Sed}^{Deg}$$
$$= \left(1.64_4\times10^9 + 1.51_2\times10^{10} + 1.37_0\times10^9 + 3.03_6\times10^7\right) mol\, Pa^{-1} h^{-1}$$
$$= 1.81_6\times10^{10}\, mol\, Pa^{-1} h^{-1}$$

The sum of all loss D-values is:

$$D^L = \sum_{i=1}^{n} D_i^{Adv} + \sum_{i=1}^{n} D_i^{Deg} = 1.60_3\times10^9\, mol\, Pa^{-1} h^{-1} + 1.81_6\times10^{10}\, mol\, Pa^{-1} h^{-1}$$
$$= 1.97_7\times10^{10}\, mol\, Pa^{-1} h^{-1}$$

The residence times associated with each of these D-values are given by $\tau = VZ/D$. For advection in individual compartments, we have:

$$\tau_A^{Adv} = \frac{V_A Z_A}{D_A^{Adv}} = \frac{1.00\times10^{14} m^3 \times 4.03\times10^{-4}\, mol\, Pa^{-1} m^{-3}}{4.03_0\times10^8\, mol\, Pa^{-1} h^{-1}} = 100._0 h$$

$$\tau_W^{Adv} = \frac{V_W Z_W}{D_W^{Adv}} = \frac{2.00\times10^{11} m^3 \times 6.00\, mol\, Pa^{-1} m^{-3}}{1.20_0\times10^9\, mol\, Pa^{-1} h^{-1}} = 1.00_0\times10^3\, h$$

$$\tau_{Sed}^{Adv} = \frac{V_{Sed} Z_{Sed}}{D_{Sed}^{Adv}} = \frac{1.00\times10^8 m^3 \times 74.4\, mol\, Pa^{-1} m^{-3}}{1.48_8\times10^5\, mol\, Pa^{-1} h^{-1}} = 5.00_0\times10^4\, h$$

For degradation in individual compartments, we have:

$$\tau_A^{Deg} = \frac{V_A Z_A}{D_A^{Deg}} = \frac{1.00\times10^{14} m^3 \times 4.03\times10^{-4}\, mol\, Pa^{-1} m^{-3}}{1.64_4\times10^9\, mol\, Pa^{-1} h^{-1}} = 24.5_1 h$$

$$\tau_W^{Deg} = \frac{V_W Z_W}{D_W^{Deg}} = \frac{2.00\times10^{11} m^3 \times 6.00\, mol\, Pa^{-1} m^{-3}}{1.51_2\times10^{10}\, mol\, Pa^{-1} h^{-1}} = 79.3_7\, h$$

$$\tau_{Soil}^{Deg} = \frac{V_{Soil} Z_{Soil}}{D_{Soil}^{Deg}} = \frac{9.00\times10^9 m^3 \times 37.3\, mol\, Pa^{-1} m^{-3}}{1.37_0\times10^9\, mol\, Pa^{-1} h^{-1}} = 245._1 h$$

$$\tau_{Sed}^{Deg} = \frac{V_{Sed} Z_{Sed}}{D_{Sed}^{Deg}} = \frac{1.00\times10^8 m^3 \times 74.4\, mol\, Pa^{-1} m^{-3}}{3.03_6\times10^7\, mol\, Pa^{-1} h^{-1}} = 245._1 h$$

The sum of VZ products appears several times in further text, so it is helpful to calculate it now:

$$\sum_{i=1}^{n} V_i Z_i = \left(1.00\times10^{14}\times4.03\times10^{-4}+2.00\times10^{11}\times6.00+9.00\times10^{9}\times37.3+1.00\times10^{8}\times74.4\right) mol\,Pa^{-1}$$

$$\sum_{i=1}^{n} V_i Z_i = \left(4.03_0\times10^{10}+1.20_0\times10^{12}+3.35_7\times10^{11}+7.44_0\times10^{9}\right) mol\,Pa^{-1}$$

$$\sum_{i=1}^{n} V_i Z_i = 1.58_3\times10^{12} mol\,Pa^{-1}$$

The residence time for loss due to all advection processes is given by:

$$\tau^{Adv} = \frac{\sum_{i=1}^{n} V_i Z_i}{\sum_{i=1}^{n} D_i^{Adv}} = \frac{1.58_3\times10^{12} mol\,Pa^{-1}}{1.60_3\times10^{9} mol\,Pa^{-1}\,h^{-1}} = 987._7\,h$$

The residence time for loss due to all degradation processes is given by:

$$\tau^{Deg} = \frac{\sum_{i=1}^{n} V_i Z_i}{\sum_{i=1}^{n} D_i^{Deg}} = \frac{1.58_3\times10^{12} mol\,Pa^{-1}}{1.81_6\times10^{10} mol\,Pa^{-1}\,h^{-1}} = 87.1_7\,h$$

The overall system residence time for all loss processes is given by:

$$\tau^{L} = \frac{\sum_{i=1}^{n} V_i Z_i}{\sum_{i=1}^{n} D_i^{Adv} + \sum_{i=1}^{n} D_i^{Deg}} = \frac{1.58_3\times10^{12} mol\,Pa^{-1}}{1.97_7\times10^{10} mol\,Pa^{-1}\,h^{-1}} = 80.1_0\,h$$

Worked Example 3.19

Assume a system fugacity of 2.024×10^{-7} *Pa* and use the *D*-values from Worked Example 3.18 to calculate the rates of each individual advective and degradation process as well as the total rates for both types of processes. Also, calculate the total rate of loss for all processes combined, and the percent contribution of each process to this total.

Recall that the rates for individual processes are given by the product of the corresponding *D*-value and the system fugacity for systems at equilibrium.

For advection, we have:

$$r_A^{Adv} = D_A^{Adv} f_{Sys} = 4.03_0\times10^{8} mol\,Pa^{-1}\,h^{-1}\times2.024\times10^{-7}\,Pa = 81.5_7\,mol\,h^{-1}$$

$$r_W^{Adv} = D_W^{Adv} f_{Sys} = 1.20_0\times10^{9} mol\,Pa^{-1}\,h^{-1}\times2.024\times10^{-7}\,Pa = 242._9\,mol\,h^{-1}$$

$$r_{Sed}^{Adv} = D_{Sed}^{Adv} f_{Sys} = 1.48_8\times10^{5} mol\,Pa^{-1}\,h^{-1}\times2.024\times10^{-7}\,Pa = 3.01_2\times10^{-2} mol\,h^{-1}$$

$$r^{T-Adv} = \left(D_A^{Adv}+D_W^{Adv}+D_{Sed}^{Adv}\right) f_{Sys} = 1.60_3\times10^{9} mol\,Pa^{-1}\,h^{-1}\times2.024\times10^{-7}\,Pa = 3.24_5\times10^{2} mol\,h^{-1}$$

For degradation, we have:

$$r_A^{Deg} = D_A^{Deg} f_{Sys} = 1.64_4 \times 10^9 mol\, Pa^{-1} h^{-1} \times 2.024 \times 10^{-7} Pa = 332._8\, mol\, h^{-1}$$

$$r_W^{Deg} = D_W^{Deg} f_{Sys} = 1.51_2 \times 10^{10} mol\, Pa^{-1} h^{-1} \times 2.024 \times 10^{-7} Pa = 3.06_0 \times 10^3 mol\, h^{-1}$$

$$r_{Soil}^{Deg} = D_{Soil}^{Deg} f_{Sys} = 1.37_0 \times 10^9 mol\, Pa^{-1} h^{-1} \times 2.024 \times 10^{-7} Pa = 277._2 mol\, h^{-1}$$

$$r_{Sed}^{Deg} = D_{Sed}^{Deg} f_{Sys} = 3.03_6 \times 10^7 mol\, Pa^{-1} h^{-1} \times 2.024 \times 10^{-7} Pa = 6.14_4 mol\, h^{-1}$$

$$r^{T-Deg} = \left(D_A^{Deg} + D_W^{Deg} + D_{Soil}^{Adv} + D_{Sed}^{Adv}\right) f_{Sys} = 1.81_6 \times 10^{10} mol\, Pa^{-1} h^{-1} \times 2.024 \times 10^{-7} Pa = 3.67_6 \times 10^3 mol\, h^{-1}$$

The total rate of loss for all processes combined is:

$$r^L = \left(D^{T-Adv} + D^{T-Deg}\right) f_{Sys} = 1.97_7 \times 10^{10} mol\, Pa^{-1} h^{-1} \times 2.024 \times 10^{-7} Pa = 4.00_1 \times 10^3 mol\, h^{-1}$$

Expressed as percentages, the contribution from each individual advection and degradation process, as well as their respective combined contributions are given by:

For advection:

$$\%r_A^{Adv} = \frac{r_A^{Adv}}{r^{T-L}} = \frac{81.5_7\, mol\, h^{-1}}{4.00_1 \times 10^3 mol\, h^{-1}} \times 100\% = 2.03_9\%$$

$$\%r_W^{Adv} = \frac{r_W^{Adv}}{r^{T-L}} = \frac{242._9\, mol\, h^{-1}}{4.00_1 \times 10^3 mol\, h^{-1}} \times 100\% = 6.07_1\%$$

$$\%r_{Sed}^{Adv} = \frac{r_{Sed}^{Adv}}{r^{L-Ov}} = \frac{3.01_2 \times 10^{-2} mol\, h^{-1}}{4.00_1 \times 10^3 mol\, h^{-1}} \times 100\% = 7.52_8 \times 10^{-4}\%$$

$$\%r^{T-Adv} = \frac{r_{Total}^{Adv}}{r^{T-L}} = \frac{3.24_5 \times 10^2 mol\, h^{-1}}{4.00_1 \times 10^3 mol\, h^{-1}} \times 100\% = 8.11_0\%$$

For degradation:

$$\%r_A^{Deg} = \frac{r_A^{Deg}}{r^{T-L}} = \frac{332._8\, mol\, h^{-1}}{4.00_1 \times 10^3 mol\, h^{-1}} \times 100\% = 8.31_8\%$$

$$\%r_W^{Deg} = \frac{r_W^{Deg}}{r^{T-L}} = \frac{3.06_0 \times 10^3 mol\, h^{-1}}{4.00_1 \times 10^3 mol\, h^{-1}} \times 100\% = 76.4_9\%$$

$$\%r_{Soil}^{Deg} = \frac{r_{Soil}^{Deg}}{r^{T-L}} = \frac{277._2 mol\, h^{-1}}{4.00_1 \times 10^3 mol\, h^{-1}} \times 100\% = 6.92_9\%$$

$$\%r_{Sed}^{Deg} = \frac{r_{Sed}^{Deg}}{r^{T-L}} = \frac{6.14_4 mol\, h^{-1}}{4.00_1 \times 10^3 mol\, h^{-1}} \times 100\% = 0.153_6\%$$

$$\%r^{T-Deg} = \frac{r^{T-Deg}}{r^{T-L}} = \frac{3.67_6 \times 10^3 mol\, h^{-1}}{4.00_1 \times 10^3 mol\, h^{-1}} \times 100\% = 91.8_9\%$$

These results show that degradation in the water is by far the most important process for loss. As well, we again see the power of the fugacity approach, in that the various *D*-values are all on the same footing, so that they can be directly compared and used in further calculations in the same manner, without requiring any conversion or scaling.

4 Non-Equilibrated Open Systems

D-Values for Diffusive Transport

Open systems that are not at equilibrium are introduced in this chapter. Most realistic environmental modelling scenarios involve interacting media compartments that are open and not at equilibrium and may or may not even be at steady state. We defer the latter scenario to Chapter 5 but consider here the former, steady state situation in which equilibrium is not present. The relationship between diffusion constants and diffusive transfer between and within compartments is developed. The rate of fugacity change is shown to be linearly dependent on the difference in fugacity between the compartments in question. The Whitman Two-Resistance approach is introduced and used to calculate the overall D-value for transfer between two compartments, based on thin interfacial films at the point of contact. The proper combination of D-values for processes that occur in series and in parallel is explained. Conversion of effective diffusivity to rates of fugacity change is derived. The very detailed algebra associated with solving for fugacities in multiple compartments not at equilibrium is developed for a four-compartment system of air, water, soil, and sediment.

4.1 DIFFUSIVE TRANSPORT WITHIN A COMPARTMENT

In practice, a chemical will move from regions of higher fugacity to those of lower fugacity by diffusion, if the two regions are in direct physical contact and movement between them is physically possible. Within a continuous single medium, diffusion will occur at a rate r^{Diff} equal to the product of a mass transfer coefficient k^M, the area over which the diffusion takes place A, and the concentration gradient ΔC:

$$r_i^{Diff} = k_i^M A \Delta C$$

The mass transfer coefficient has units of $m\ h^{-1}$ and can be considered a "diffusion velocity". Given area in units of m^2 and concentration in $mol\ m^{-3}$, we arrive at r^{Diff} in our preferred units of $mol\ h^{-1}$.

To express this relation in terms of fugacity, we need to recognize that $\Delta C = \Delta(Zf) = Z\Delta f$:

$$r_i^{Diff} = k_i^M A \Delta\left(Zf_i\right) = k_i^M A Z_i \Delta f_i$$

If we define a D-value for diffusion within a compartment as:

$$\boxed{D_i^{Diff} = k_i^M A Z_i}$$

We can then write a simple expression for diffusive flow within a compartment as:

$$\boxed{r_i^{Diff} = D_i^{Diff} \Delta f_i}$$

DOI: 10.1201/9781003657170-4

Worked Example 4.1

A thin film of water separates a medium with fugacity $f_A = 2.0\ Pa$ from another medium with fugacity $f_B = 1.6\ Pa$. The chemical's fugacity capacity in water is $Z_W = 5.0\ mol\ Pa^{-1}m^{-3}$. What is the rate of diffusion across 10 m^2 of this thin film, if the mass-transfer coefficient for diffusion in the water is 2.0 $m\ h^{-1}$?

Here the D-value for diffusion is given by:

$$D_W^{Diff} = k_W^M A Z_W$$

$$D_W^{Diff} = 2.0\,m\,h^{-1} \times 10.\,m^2 \times 5.0\,mol\,Pa^{-1}m^{-3}$$

$$D_W^{Diff} = 1.0_0 \times 10^2 mol\,Pa^{-1}h^{-1}$$

The diffusive flow rate is given by:

$$r_W^{Diff} = D_W^{Diff}\,\Delta f_W$$

$$r_W^{Diff} = 1.0_0 \times 10^2\,mol\,Pa^{-1}h^{-1} \times (2.0\,Pa - 1.6\,Pa)$$

$$r_W^{Diff} = 40.\,mol\,h^{-1}$$

The problem of determining evaporation loss rates is very similar. The assumption is made that the fugacity at the evaporative surface is equal to the product of the evaporating substance's vapour pressure and its mole fraction (Raoult's law condition), and that the fugacity of the surrounding air is a lower background value. Evaporation occurs as a result of diffusion through a thin layer of air above the liquid surface which has a fugacity gradient equal to the difference between the high fugacity at the liquid's surface and the lower background fugacity in the surrounding air.

Under these assumptions, the rate of evaporative loss is equal to the rate of diffusion across this surface gradient, given as:

$$r_i^{Diff} = D_i^{Diff}\,\Delta f_i$$

where

$$D_i^{Diff} = k_i^M A Z_i$$

$$\Delta f_i = f_{Surface} - f_{Background}$$

Worked Example 4.2

A shallow container filled with water with a surface area of 0.17 m^2 is left standing in a room with a background water concentration in air of 0.15 $mol\ m^{-3}$. Given that the vapour pressure at the water surface is $4.25 \times 10^3\ Pa$ and using a mass transfer coefficient for water diffusion of 50 $m\ h^{-1}$, calculate the rate of evaporation of the water.

To determine the rate of evaporation, we need to determine both Z_A as well as the fugacity of water in the background air. Since the temperature is not specified, we shall assume as usual that $Z_A = 4.0 \times 10^{-4}\ mol\ Pa^{-1}\ m^{-3}$.

The background fugacity of water in air is given by:

$$f_{Background} = \frac{C_{Background}}{Z_A}$$

$$f_{Background} = \frac{0.15\,mol\,m^{-3}}{4.0 \times 10^{-4}\,mol\,Pa^{-1}\,m^{-3}}$$

$$f_{Background} = 3.7_5 \times 10^2\,Pa$$

Since chemical's fugacity in the gas phase is essentially equal to the partial pressure of a chemical at pressures typical of environmental systems, we assume that the surface fugacity is equal to the partial pressure at the liquid's surface (the vapour pressure):

$$f_{Surface} = 4.25 \times 10^3\,Pa$$

Thus, the fugacity difference across the air layer is:

$$\Delta f_A = f_{Surface} - f_{Background}$$

$$\Delta f_A = 4.25 \times 10^3\,Pa - 3.7_5 \times 10^2\,Pa$$

$$\Delta f_A = 3.87_5 \times 10^3\,Pa$$

The D-value for diffusion through the air layer is given by:

$$D_A^{Diff} = k_A^M A Z_A$$

$$D_A^{Diff} = 50.\ m\ h^{-1} \times 0.17\,m^2 \times 4.0 \times 10^{-4}\,mol\,Pa^{-1}m^{-3}$$

$$D_A^{Diff} = 3.4_0 \times 10^{-3}\,mol\,Pa^{-1}h^{-1}$$

Finally, the rate of evaporation by diffusion is given by:

$$r_A^{Diff} = D_A^{Diff}\Delta f_A$$

$$r_A^{Diff} = 3.4_0 \times 10^{-3}\,mol\,Pa^{-1}h^{-1} \times 3.87_5 \times 10^3\,Pa$$

$$r_A^{Diff} = 13\,mol\,h^{-1}$$

Some diffusion processes are viewed in terms of molecular diffusivity B ($m^2\,h^{-1}$), instead of mass transfer coefficients. The two are related through the distance over which the diffusion is occurring, Y (m):

$$k_i^M = \frac{B_i}{Y_i}$$

Worked Example 4.3

The molecular diffusivity of a chemical in air is $1.80 \times 10^{-2}\ m^2\,h^{-1}$. Assuming diffusion is occurring across a $4.75 \times 10^{-3}\ m$ distance, calculate the mass transfer coefficient for this diffusion.

The calculation is:

$$k_A^M = \frac{B_A}{Y_A}$$

$$k_A^M = \frac{1.80 \times 10^{-2}\,m^2h^{-1}}{4.75 \times 10^{-3}\,m}$$

$$k_A^M = 3.79\,m\,h^{-1}$$

4.2 DIFFUSIVE TRANSPORT BETWEEN COMPARTMENTS

It is often the case that environmental compartments in direct contact are not at the same fugacity with respect to a contaminant. In this case, the fugacity difference or gradient becomes a driving force for the chemical to move by diffusion from the compartment with higher fugacity to the compartment with lower fugacity. In closed systems where there is no advection or degradation, this fugacity-driven diffusive movement would eventually result in the fugacity in all compartments coming to the same value (equifugacity), at which point the system would achieve equilibrium. However, in open systems, where advection and degradation occur, it is common for the fugacity-driven diffusive flow of a chemical to be offset by losses from the lower-fugacity compartment, such that equilibrium is never achieved. In this case, we have a non-equilibrium, steady-state system, in which two or more compartments have different but unchanging fugacities and there is a constant rate of diffusive transfer between compartments where exchange of chemical is physically possible. Such scenarios are termed "Level III" in the Mackay terminology.

The basic relationship that is used in this situation is one which relates the rate of diffusive transfer between two compartments in contact to an overall D-value for diffusion and the difference in fugacity between the two compartments:

$$r_{ij}^{Diff} = D_{ij}^{Diff-Ov}\left(f_i - f_j\right)$$

Here, the direction of the diffusive transport is determined by the sign of the difference in fugacity. For example, if $f_i > f_j$, the transport will be from compartment "i" to compartment "j". Figure 4.1 shows the direction of diffusive intercompartment transport for three scenarios involving different relative fugacity magnitudes between air and water in contact.

The key to solving steady-state problems at disequilibrium is in setting up mass balance expressions for each compartment in which the rates of input equal the rates of output, such that there is no change in mass with time. This is followed by a determination of the fugacity in each compartment, and subsequent determinations of concentrations and amounts as introduced in the previous chapters.

The only new D-value that we need to develop is the "overall" diffusion D-value which applies to diffusion across the interface between two compartments for which diffusive transfer can occur. This overall D-value is analogous to a conductivity in electrical theory. It is deduced using the Whitman Two-Resistance mass transfer coefficient approach, expressed in terms of a pair of D-values in series associated with diffusive transport through a thin layer of medium on either side of the interfacial boundary in question. The reciprocal of the overall D-value is analogous to electrical resistance, and as a result is sometimes itself referred to as a resistance.

In general, the reciprocal of the overall D-value is given by the sum of the reciprocal of the D-values for the two compartments between which the diffusion occurs:

$$\boxed{\frac{1}{D_{ij}^{Diff-Ov}} = \frac{1}{D_{ii}^{Diff-i}} + \frac{1}{D_{ij}^{Diff-j}}}$$

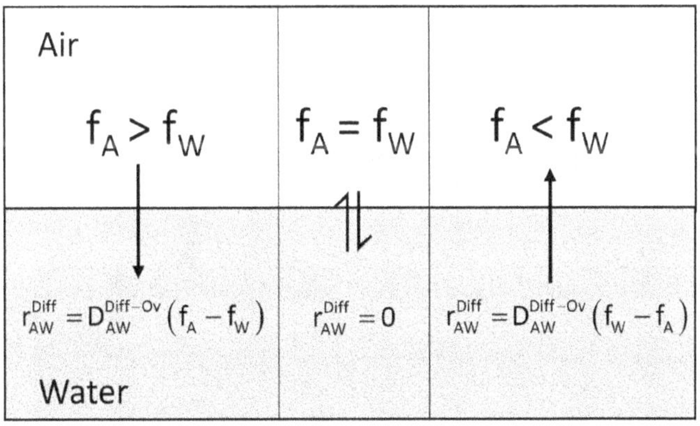

FIGURE 4.1 Direction of diffusive, fugacity-gradient-driven, intercompartment transport between air and water under three scenarios with different relative fugacities for a thin layer of each respective compartment.

Note that the nomenclature used implies that the individual D-values are specific both to the compartment in question and to the two compartments on either side of the interface. Thus, the D-value for air over water D_{AW}^{Diff-A} may be different from the D-value for air over octanol D_{AO}^{Diff-A}, and certainly will be so for air over soil D_{A-Soil}^{Diff-A}, due to differences in the thickness and overall structure of the interfacial boundary layer. Indeed, the value of D_{AW}^{Diff-A} will even vary between different bodies of water due to differences in wave action associated with weather conditions, which alter the nature of the interfacial layer.

In order to avoid confusion, all D-values that pertain to processes that involve transfer between two compartments are denoted with corresponding subscripts listed in alphabetical order. This choice is made since diffusive transfer may occur in either direction and is modelled with the same D-value.

Practically, the reciprocal relationship given above results in the following rather awkward-looking expression for the overall D-value itself:

$$\boxed{D_{ij}^{Diff-Ov} = \frac{1}{\left(\dfrac{1}{D_{ij}^{Diff-i}} + \dfrac{1}{D_{ij}^{Diff-j}}\right)}}$$

The magnitude of the overall D-value is usually limited by the smallest D-value in this expression. In such cases, the overall diffusion is said to be controlled by the diffusion on the "side" of the interfacial boundary with the smallest D-value. For example, if $D_{AW}^{Diff-A} << D_{AW}^{Diff-W}$, the result will be that $D_{AW}^{Diff-Ov} \simeq D_{AW}^{Diff-A}$ and we would say that the process is "air-side controlled". Here, the rate of diffusion is limited by the rate at which the chemical passes through the thin air layer at the interface, as the rate of diffusion through the corresponding water layer is much faster and therefore less important, and thus non-rate-limiting. This relationship is easily deduced in the following way for the circumstance where $D_{AW}^{Diff-A} << D_{AW}^{Diff-W}$:

$$D_{AW}^{Diff-Ov} = \frac{1}{\left(\dfrac{1}{D_{AW}^{Diff-A}} + \dfrac{1}{D_{AW}^{Diff-W}}\right)} \simeq \frac{1}{\left(\dfrac{1}{D_{AW}^{Diff-A}}\right)} = D_{AW}^{Diff-A}$$

Each of the two D-values specific to the compartments between which diffusion is occurring is given by the same type of expression as introduced above:

$$D_{ij}^{Diff-i} = k_{ij}^{M-i} A_{ij} Z_i$$

Individual mass transfer coefficients employed in the Whitman Two-Resistance model are named by the "side" of the intermedia boundary interface to which they pertain. For example, the overall D-value for an air–water interface $D_{AW}^{Diff-Ov}$ would involve k_{AW}^{M-A}, the "air-side" mass-transfer coefficient and k_{AW}^{M-W}, the "water-side" mass transfer coefficient.

As an example, the overall D-value for diffusive transfer of a chemical between air and water would be defined in the Whitman Two-Resistance approach as:

$$\frac{1}{D_{AW}^{Diff-Ov}} = \frac{1}{D_{AW}^{Diff-A}} + \frac{1}{D_{AW}^{Diff-W}} = \frac{1}{k_{AW}^{M-A} A_{AW} Z_A} + \frac{1}{k_{AW}^{M-W} A_{AW} Z_W}$$

and

$$D_{AW}^{Diff-Ov} = \frac{1}{\left(\dfrac{1}{D_{AW}^{Diff-A}} + \dfrac{1}{D_{AW}^{Diff-W}}\right)} = \frac{1}{\left(\dfrac{1}{k_{AW}^{M-A} A_{AW} Z_A} + \dfrac{1}{k_{AW}^{M-W} A_{AW} Z_W}\right)}$$

Worked Example 4.4

The EQC environment values for the air-side and water-side mass transfer coefficients for air–water diffusive transport are 5.0 $m\ h^{-1}$ and 5.0 × 10^{-2} $m\ h^{-1}$, respectively. Use these values to calculate the overall D-value for diffusive transport at the air–water interface over a 1.0 × 10^{10} m^2 area. Use a Henry's Law constant of 1.67 × 10^{-1} $m^3\ Pa\ mol^{-1}$ and assume $Z_A = 4.03 \times 10^{-4}\ mol\ Pa^{-1}\ m^{-3}$.

The fugacity capacity in water is given by $1/H$:

$$Z_W = \frac{1}{H} = \frac{1}{1.67 \times 10^{-1} m^3 Pa\ mol^{-1}} = 5.98_8\ mol\ Pa^{-1} m^{-3}$$

The individual D-values are:

$$D_{AW}^{Diff-A} = k_{AW}^{M-A} A_{AW} Z_A$$

$$D_{AW}^{Diff-A} = 5.0\ m\ h^{-1} \times 1.0 \times 10^{10} m^2 \times 4.03 \times 10^{-4}\ mol\ Pa^{-1} m^{-3}$$

$$D_{AW}^{Diff-A} = 2.0_2 \times 10^7\ mol\ Pa^{-1} h^{-1}$$

$$D_{AW}^{Diff-W} = k_{AW}^{M-W} A_{AW} Z_W$$

$$D_{AW}^{Diff-W} = 5.0 \times 10^{-2}\ m\ h^{-1} \times 1.0 \times 10^{10} m^2 \times 5.98_8\ mol\ Pa^{-1} m^{-3}$$

$$D_{AW}^{Diff-W} = 2.9_9 \times 10^9\ mol\ Pa^{-1} h^{-1}$$

The overall D-value for interfacial diffusive transport across the air–water interface is:

$$D_{AW}^{Diff-Ov} = \frac{1}{\left(\frac{1}{D_{AW}^{Diff-A}} + \frac{1}{D_{AW}^{Diff-W}}\right)}$$

$$D_{AW}^{Diff-Ov} = \frac{1}{\left(\frac{1}{2.0_2 \times 10^7\ mol\ Pa^{-1} h^{-1}} + \frac{1}{2.9_9 \times 10^9\ mol\ Pa^{-1} h^{-1}}\right)}$$

$$D_{AW}^{Diff-Ov} = \frac{1}{\left(4.9_6 \times 10^{-8} mol\ Pa^{-1} h^{-1} + 3.3_4 \times 10^{-10}\ mol\ Pa^{-1} h^{-1}\right)}$$

$$D_{AW}^{Diff-Ov} = 2.0_0 \times 10^7\ mol\ Pa^{-1}\ h^{-1}$$

Note that the overall D-value for diffusion at the air–water interface is nearly identical to the air-side D-value. Here, as discussed above, the diffusion is said to be "air-side controlled".

4.3 COMBINING TRANSPORT PROCESSES: VOLATILIZATION FROM SOILS

In most circumstances of interest in environmental modelling, there will exist more than one transport process that involves movement of a chemical from one environmental compartment to another. For example, a chemical may leave an air body to enter a water body by way of air–water interfacial diffusion, dry aerosol deposition, and rain deposition, including wet deposition via aerosol washout. The extra work of developing dexterity with D-values pays off in such circumstances, as D-values for these various processes are easily combined to give an overall D-value for intercompartment transfer. Such combination approaches are directly analogous to electrical resistance theory, as will be evident to anyone familiar with that theory.

There are two ways in which such D-values are combined. If the processes in question occur in parallel, i.e., as separate processes from the same origin whose rates do not depend on one another, the corresponding D-values simply add. So, for parallel processes $P1$, $P2$, $P3$, etc., between phases i and j, we have:

$$D_{ij}^{T} = D_{ij}^{P1} + D_{ij}^{P2} + D_{ij}^{P3} + \ldots$$

For processes that occur in series $D_{ij}^{S(n)}$, such as intermedia diffusive transport introduced above, the D-values add reciprocally, as:

$$\frac{1}{D_{ij}^{T}} = \frac{1}{D_{ij}^{S1}} + \frac{1}{D_{ij}^{S2}} + \frac{1}{D_{ij}^{S3}} + \ldots$$

Circumstances also arise where both parallel and series transport processes are active, in which case D-values are combined using both approaches. Suppose we have three transport processes, two of which are parallel ($P1$ and $P2$) and one of which is in series ($S1$) with both parallel processes. The overall D-value for the combination of all three of these processes would be:

$$\frac{1}{D_{ij}^{T}} = \frac{1}{D_{ij}^{S1}} + \frac{1}{\left(D_{ij}^{P1} + D_{ij}^{P2}\right)}$$

or

$$D_{ij}^{T} = \frac{1}{\left(\dfrac{1}{D_{ij}^{S1}} + \dfrac{1}{\left(D_{ij}^{P1} + D_{ij}^{P2}\right)}\right)}$$

Alternatively, we may have three transport processes, one of which ($P1$) is parallel to the other two, and two of which ($S1$ and $S2$) are in series with each other, and mutually parallel to the first. The overall D-value for the combination of all three of these processes would be:

$$D_{ij}^{T} = D_{ij}^{P1} + \left[\frac{1}{\left(\dfrac{1}{D_{ij}^{S1} + D_{ij}^{S2}}\right)}\right]$$

Figure 4.2 illustrates these two scenarios graphically.

Chemical transfer between soil and air is an excellent example in which D-values for transport processes are combined both as parallel and series processes to obtain an overall D-value for soil–air transfer. Such a process is treated in the manner of Jury et al. (1983), in which the transfer is assumed to be due to diffusion through pore air and pore water of soil as parallel processes, and diffusion through a thin air boundary layer at the soil surface as a series process.

Together, these combine to give an overall soil–air diffusive transfer D-value as follows:

$$\boxed{\frac{1}{D_{A-Soil}^{Diff-Ov}} = \frac{1}{D_{A-Soil}^{A-Boundary}} + \frac{1}{\left(D_{A-Soil}^{PoreA} + D_{Soil-W}^{PoreW}\right)}}$$

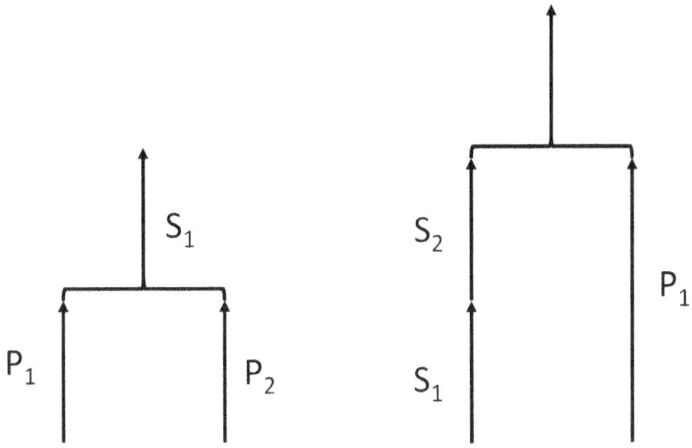

$$D^{T}_{ij} = \frac{1}{\left[\dfrac{1}{D^{S_1}_{ij}} + \dfrac{1}{\left(D^{P_1}_{ij} + D^{P_2}_{ij}\right)}\right]} \qquad D^{T}_{ij} = D^{P_1}_{ij} + \frac{1}{\left[\dfrac{1}{\left(D^{S_1}_{ij} + D^{S_2}_{ij}\right)}\right]}$$

FIGURE 4.2 Two scenarios illustrating the manner in which overall D-values are calculated for two commonly found combinations of parallel and series processes. The left image depicts two parallel processes followed by a single process in series. The right depicts two series processes parallel to a third process. In each case, the total D-value comprises combinations of individual D-values for the series and parallel steps.

Worked Example 4.5

A chemical moves through the pore air and pore water of soil by parallel diffusion processes, then passes through the air boundary layer in a process in series with the first two processes. Given the following D-values for each of these processes, determine the overall D-value for volatilization through the soil: $D^{PoreA}_{A-Soil} = 5.00 \times 10^{-2}\ mol\ Pa^{-1}\ h^{-1}$, $D^{PoreW}_{Soil-W} = 1.00 \times 10^{-2}\ mol\ Pa^{-1}\ h^{-1}$ and $D^{A-Boundary}_{A-Soil} = 20.0\ mol\ Pa^{-1}\ h^{-1}$

As introduced above, these three process D-values combine according to:

$$\frac{1}{D^{Diff-Ov}_{A-Soil}} = \frac{1}{D^{Air-Boundary}_{A-Soil}} + \frac{1}{\left(D^{PoreA}_{A-Soil} + D^{PoreW}_{Soil-W}\right)}$$

$$\frac{1}{D^{Diff-Ov}_{A-Soil}} = \frac{1}{20.0\ mol\ Pa^{-1}h^{-1}} + \frac{1}{\left(5.00 \times 10^{-2} mol\ Pa^{-1}h^{-1} + 1.00 \times 10^{-2} mol\ Pa^{-1}h^{-1}\right)}$$

$$\frac{1}{D^{Diff-Ov}_{A-Soil}} = \frac{1}{20.0\ mol\ Pa^{-1}h^{-1}} + \frac{1}{6.00 \times 10^{-2} mol\ Pa^{-1}h^{-1}} = 16.6_7\ Pa\,h\,mol^{-1}$$

$$D^{Diff-Ov}_{A-Soil} = 5.98 \times 10^{-2} mol\ Pa^{-1}\ h^{-1}$$

Note that the overall D-value for soil–air transfer is essentially the same as that of the combined pore air and water values for the soil, as these are by far the slowest and are rate-limiting for the overall process.

Diffusion in soils is often modelled using the Millington–Quirk (MQ) expression for effective diffusivity in pore air and pore water as a function of the air and water content of the soil. In this approach, effective diffusivities are estimated by expressions incorporating the soil–air and soil–water volume fractions, v_{Soil}^{f-A} and v_{Soil}^{f-W}:

$$B_{Soil}^{Eff-A} = \frac{B_{Soil}^{A}\left(v_{Soil}^{f-A}\right)^{10/3}}{\left(v_{Soil}^{f-A} + v_{Soil}^{f-W}\right)^{2}}$$

$$B_{Soil}^{Eff-W} = \frac{B_{Soil}^{W}\left(v_{Soil}^{f-W}\right)^{10/3}}{\left(v_{Soil}^{f-A} + v_{Soil}^{f-W}\right)^{2}}$$

With these effective diffusivities in hand, the mass transfer coefficients for both processes can be estimated, and the corresponding D-values can be evaluated.

The D-value for diffusion through the air boundary layer at the soil–air interface can also be expressed in terms of molecular diffusivity, but since the diffusion is through pure air, there is no need for any correction:

$$D_{A-Soil}^{A-Boundary} = \frac{B_{A-Soil}^{A}\,A_{A-Soil}\,Z_{A}}{Y_{A-Soil}^{Boundary}}$$

D-values for diffusion in the soil pore air and pore water incorporate the effective diffusivities introduced above, and are given by:

$$D_{A-Soil}^{PoreA} = \frac{B_{Soil}^{Eff-A}\,A_{A-Soil}\,Z_{A}}{Y_{Soil}}$$

and

$$D_{Soil-W}^{PoreW} = \frac{B_{Soil}^{Eff-W}\,A_{A-Soil}\,Z_{W}}{Y_{Soil}}$$

The following example uses input data for DDT in a carrot field from the *Soil* model available from the CEMC website (https://www.trentu.ca/cemc/resources-and-models).

Worked Example 4.6

DDT volatilization from a $1.00 \times 10^4\ m^2$ field with soil comprising 20.0% air and 30.0% water passes through an air boundary layer of thickness $4.75 \times 10^{-3}\ m$. The molecular diffusivity in soil–air and soil–water are determined to be $0.430\ m^2\ day^{-1}$ and $4.30 \times 10^{-5}\ m^2\ day^{-1}$, respectively. Use a diffusion path length of 5.00 cm in soil, $Z_A = 4.03 \times 10^{-4}\ mol\ Pa^{-1}\ m^{-3}$ and $Z_w = 4.37 \times 10^{-1}\ mol\ Pa^{-1}\ m^{-3}$ to compute the overall D-value for volatilization of DDT from the soil.

To calculate the overall D-value for volatilization, we must determine the D-values for the three processes of diffusion through soil pore air, soil pore water, and through the soil–air boundary layer. The first two require adjusting the molecular diffusivities for the properties of the soil.

The effective diffusivity in soil pore air is:

$$B_{Soil}^{Eff-A} = \frac{B_{Soil}^{A}\left(v_{Soil}^{f-A}\right)^{10/3}}{\left(v_{Soil}^{f-A} + v_{Soil}^{f-W}\right)^{2}}$$

$$B_{Soil}^{Eff-A} = \frac{\left(0.430\ m^2day^{-1}/24\,h\,day^{-1}\right)(0.200)^{10/3}}{(0.200 + 0.300)^{2}}$$

$$B_{Soil}^{Eff-A} = 3.35_3 \times 10^{-4}\,m^2 h^{-1}$$

The effective molecular diffusivity in soil pore water is:

$$B_{Soil}^{Eff-W} = \frac{B_{Soil}^{W}\left(v_{Soil}^{f-W}\right)^{10/3}}{\left(v_{Soil}^{f-A} + v_{Soil}^{f-W}\right)^{2}}$$

$$B_{Soil}^{Eff-W} = \frac{\left(4.30\times10^{-5}\,m^2 day^{-1}/24\,h\,day^{-1}\right)(0.300)^{10/3}}{(0.200+0.300)^2}$$

$$B_{Soil}^{Eff-W} = 1.29_5\times10^{-7}\,m^2h^{-1}$$

The three *D*-values associated with the various diffusion processes may now be calculated. For the soil–air interfacial air boundary layer we have:

$$D_{Soil-A}^{A-Boundary} = \frac{B_{Soil-A}^{A} A_{Soil} Z_A}{Y_{Soil}}$$

$$D_{A-Soil}^{A-Boundary} = \frac{\left(0.430\,m^2 day^{-1}/24\,h\,day^{-1}\right)\times1.00\times10^4 m^2\times4.03\times10^{-4} mol\,Pa^{-1}m^{-3}}{4.75\times10^{-3} m}$$

$$D_{A-Soil}^{A-Boundary} = 15.2_0\,mol\,Pa^{-1}h^{-1}$$

The *D*-value for diffusion through pore air in the soil is:

$$D_{A-Soil}^{PoreA} = \frac{B_{Soil}^{Eff-A} A_{Soil} Z_A}{Y_{Soil}}$$

$$D_{A-Soil}^{PoreA} = \frac{3.35_3\times10^{-4} m^2h^{-1}\times1.00\times10^4 m^2\times4.03\times10^{-4} mol\,Pa^{-1}m^{-3}}{\left(5.00\,cm/100\,cm\,m^{-1}\right)}$$

$$D_{A-Soil}^{PoreA} = 2.70_2\times10^{-2} mol\,Pa^{-1}h^{-1}$$

The *D*-value for diffusion through pore water in the soil is:

$$D_{Soil-W}^{PoreW} = \frac{B_{Soil}^{Eff-W} A_{Soil} Z_W}{Y_{Soil}}$$

$$D_{Soil-W}^{PoreW} = \frac{1.29_5\times10^{-7} m^2h^{-1}\times1.00\times10^4 m^2\times4.37\times10^{-1} mol\,Pa^{-1}m^{-3}}{\left(5.00\,cm/100\,cm\,m^{-1}\right)}$$

$$D_{Soil-W}^{PoreW} = 1.13_2\times10^{-2} mol\,Pa^{-1}h^{-1}$$

With these three *D*-values in hand, we can combine them according to the equation introduced above for two parallel processes occurring in series with a third process:

$$\frac{1}{D_{A-Soil}^{Diff-Ov}} = \frac{1}{D_{A-Soil}^{Air-Boundary}} + \frac{1}{\left(D_{A-Soil}^{PoreA} + D_{Soil-W}^{PoreW}\right)}$$

$$\frac{1}{D_{A-Soil}^{Diff-Ov}} = \frac{1}{15.2_0\ mol\ Pa^{-1}h^{-1}} + \frac{1}{\left(2.70_2\times10^{-2} mol\ Pa^{-1}h^{-1} + 1.13_2\times10^{-2} mol\ Pa^{-1}h^{-1}\right)}$$

$$\frac{1}{D_{A-Soil}^{Diff-Ov}} = \frac{1}{15.2_0\ mol\ Pa^{-1}h^{-1}} + \frac{1}{3.83_5\times10^{-2} mol\ Pa^{-1}h^{-1}} = 26.1_4\,Pa\,h\,mol^{-1}$$

$$D_{A-Soil}^{Diff-Ov} = 3.82\times10^{-2} mol\,Pa^{-1}h^{-1}$$

As in the previous Worked Example, the overall *D*-value for volatilization is essentially the same as that of the combined soil pore air and pore water values, as these are rate limiting for the overall process.

4.4 FUGACITY DETERMINATION IN OPEN SYSTEMS NOT AT EQUILIBRIUM

Unlike systems at equilibrium, which are characterized by a single prevailing fugacity, systems that are not at equilibrium have different fugacities in each compartment. As mentioned previously, such situations require a treatment termed Level III in the Mackay terminology, if they are at steady state. In the event that the concentrations of the chemical in question are *known* in each compartment, determination of the corresponding fugacities is a relatively straightforward process involving estimation of the individual or bulk fugacity capacity for each compartment, as appropriate:

$$f_i = \frac{C_i}{Z_i} \text{ or } f_i = \frac{C_i}{Z_i^{Bulk}}$$

From these fugacities, the various rates of transport between compartments can be directly determined by the product of the *D*-value for the process and the appropriate fugacity:

$$r_i = D_i f_i$$

Worked Example 4.7

The following table provides *D*-values for various processes involved in sediment–water DDT exchange. Use these *D*-values and the appropriate fugacity to determine the rates for each process, the total rate of transfer from water to sediment, and the total rate of transfer from sediment to water. Assume DDT concentrations of 2.82×10^{-7} *mol* m^{-3} in the bulk water phase, 1.24×10^{-2} *mol* m^{-3} in the bulk sediment phase, and bulk fugacity capacities of 7.19×10^{-1} *mol* Pa^{-1} m^{-3} and 4.976×10^{3} *mol* Pa^{-1} m^{-3}.

Transport Processes	*D*-value (*mol* Pa^{-1} h^{-1})
Overall sediment–water diffusion	4.17×10^{-2}
Deposition	7.05
Resuspension	0.470
Burial	0.705
Degradation in sediment	0.0862

To start, we need to determine the bulk water and bulk sediment fugacities from *C/Z:*

$$f_W = \frac{C_W^{Bulk}}{Z_W^{Bulk}} = \frac{2.82 \times 10^{-7}\, mol\, m^{-3}}{7.19 \times 10^{-1}\, mol\, Pa^{-1} m^{-3}} = 3.92_2 \times 10^{-7}\, Pa$$

$$f_{Sed} = \frac{C_{Sed}^{Bulk}}{Z_{Sed}^{Bulk}} = \frac{1.24 \times 10^{-2}\, mol\, m^{-3}}{4.976 \times 10^{3}\, mol\, Pa^{-1} m^{-3}} = 2.49_2 \times 10^{-6}\, Pa$$

Now the rates of the various processes follow directly from the product *Df:*

$$r_{Sed-W}^{Diff-W} = D_{Sed-W}^{Diff-Ov} f_W = 4.17 \times 10^{-2} mol\, Pa^{-1} h^{-1} \times 3.92_2 \times 10^{-7}\, Pa = 1.63_6 \times 10^{-8} mol\, h^{-1}$$

$$r_{Sed-W}^{Dep} = D_{Sed-W}^{Dep} f_W = 7.05 mol\, Pa^{-1} h^{-1} \times 3.92_2 \times 10^{-7}\, Pa = 2.76_5 \times 10^{-6} mol\, h^{-1}$$

$$r_{Sed-W}^{Diff-Sed} = D_{Sed-W}^{Diff-Ov} f_{Sed} = 4.17 \times 10^{-2} mol\, Pa^{-1} h^{-1} \times 2.49_2 \times 10^{-6}\, Pa = 1.03_9 \times 10^{-7} mol\, h^{-1}$$

$$r_{Sed-W}^{Resusp} = D_{Sed-W}^{Resusp} f_{Sed} = 4.70 \times 10^{-1} mol\, Pa^{-1} h^{-1} \times 2.49_2 \times 10^{-6}\, Pa = 1.17_1 \times 10^{-6} mol\, h^{-1}$$

$$r_{Sed}^{Burial} = D_{Sed}^{Burial} f_{Sed} = 7.05 \times 10^{-1} mol\, Pa^{-1} h^{-1} \times 2.49_2 \times 10^{-6}\, Pa = 1.75_7 \times 10^{-6} mol\, h^{-1}$$

$$r_{Sed}^{Deg} = D_{Sed}^{Deg} f_{Sed} = 8.62 \times 10^{-2} mol\, Pa^{-1} h^{-1} \times 2.49_2 \times 10^{-6}\, Pa = 2.14_8 \times 10^{-7} mol\, h^{-1}$$

The total rate of transfer from water to sediment is:

$$r_{Sed-W}^{T-W} = r_{Sed-W}^{Diff-W} + r_{Sed-W}^{Dep} = 1.63_6 \times 10^{-8} mol\,h^{-1} + 2.76_5 \times 10^{-6} mol\,h^{-1} = 2.78_1 \times 10^{-6} mol\,h^{-1}$$

The total rate of transfer from sediment to water is:

$$r_{Sed-W}^{T-Sed} = r_{Sed-W}^{Diff-Sed} + r_{Sed-W}^{Resusp} = 1.03_9 \times 10^{-7} mol\,h^{-1} + 1.17_1 \times 10^{-6} mol\,h^{-1} = 1.27_5 \times 10^{-6} mol\,h^{-1}$$

The net rate of transfer is from water to sediment:

$$r_{Sed-W}^{Net} = r_{Sed-W}^{T-W} - r_{Sed-W}^{T-Sed} = 2.78_1 \times 10^{-6} mol\,h^{-1} - 1.27_5 \times 10^{-6} mol\,h^{-1} = 1.50_6 \times 10^{-6} mol\,h^{-1}$$

Thus, the net transfer is from water to sediment, even though the fugacity in the sediment is higher than in the water. This is due to the fact that the deposition rate is higher than the resuspension rates. As well, the deposition rates are much higher than the diffusion rates and dominate the transfer process.

The determination of compartment fugacities in non-equilibrium steady-state systems, when the compartment concentrations are *unknown*, is a very different and much more mathematically challenging task. This circumstance is commonly encountered in modelling emissions for which a model environment is constructed, and a chemical is "released" into one or more compartments. In this case, it is necessary to develop a series of mass balance equations, one per compartment, with as many unknown fugacities. Once these equations have been constructed, algebraic or other types of methods must be used to solve for the individual fugacities.

For example, consider the sustained release of a chemical into the water of a two-compartment model environment consisting of air and water in contact, at steady state. Ignoring advection for relative simplicity (which also implies a negligible volume change from the chemical's introduction), transport processes may include diffusion between air and water, as well as chemical transfer from air to water via dry and wet deposition. We require two independent equations that are functions of f_A and f_W.

A highly systematic approach is to develop a set of mass balance equations for the rate of input and output for each compartment. We can write the following expression for the water compartment at steady state:

$$\sum_{i=1}^{all\,sources} r_W^i = \sum_{j=1}^{all\,losses} r_W^j$$

The sources here include the direct emission into water, diffusive transfer from air to water, and dry and, wet deposition. Losses include degradation in water and diffusive transfer from water to air.

Therefore, we can write a detailed mass balance statement for water at steady state as:

$$E_W + r_{AW}^{Diff-A} + r_A^{Dry} + r_A^{Wet} + r_A^{Rain} = r_{AW}^{Diff-W} + r_W^{Deg}$$

Expressed in terms of D-values and the appropriate fugacities, we have:

$$E_W + \left(D_{AW}^{Diff-Ov} + D_A^{Dry} + D_A^{Wet} + D_A^{Rain}\right) f_A = \left(D_{AW}^{Diff-Ov} + D_W^{Deg}\right) f_W$$

We can develop a similar mass balance for air at steady state in terms of rates and in terms of D-values with fugacities:

$$r_{AW}^{Diff-W} = r_{AW}^{Diff-A} + r_A^{Dry} + r_A^{Wet} + r_A^{Rain} + r_A^{Deg}$$

$$D_{AW}^{Diff-Ov} f_W = \left(D_{AW}^{Diff-Ov} + D_A^{Dry} + D_A^{Wet} + D_A^{Rain} + D_A^{Deg}\right) f_A$$

The subsequent determination of f_W and f_A will be much clearer if we represent the sum of all loss processes in the water and in the air as:

$$D_W^L = \left(D_{AW}^{Diff-Ov} + D_W^{Deg}\right)$$

$$D_A^L = \left(D_{AW}^{Diff-Ov} + D_A^{Dry} + D_A^{Wet} + D_A^{Rain} + D_A^{Deg}\right)$$

These loss processes are illustrated schematically in Figure 4.3.

As well, the total of all processes that transfer chemical from air to water can be represented by:

$$D_{AW}^{T-A} = \left(D_{AW}^{Diff-Ov} + D_A^{Dry} + D_A^{Wet} + D_A^{Rain}\right)$$

There is no value in an analogous equation for transfer from water to air, since there is only one process that goes in this direction, namely diffusive transfer from water to air.

With these condensed symbols, the water and air mass balances are now:

$$E_W + D_{AW}^{T-A} f_A = D_W^L f_W$$

$$D_{AW}^{Diff-Ov} f_W = D_A^L f_A$$

Isolating f_W from the first equation, we have:

$$f_W = \frac{E_W + D_{AW}^{T-A} f_A}{D_W^L}$$

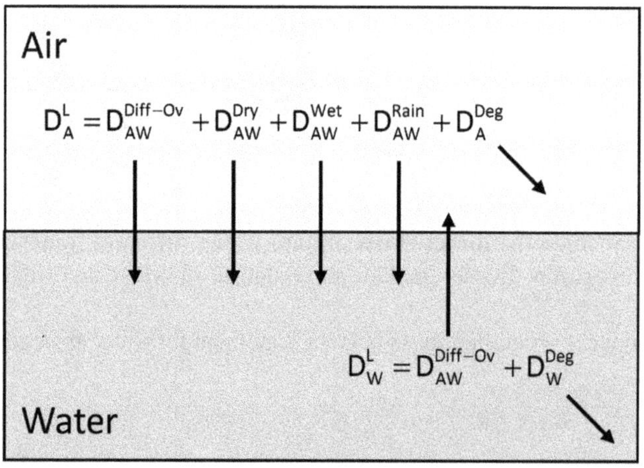

FIGURE 4.3 Loss mechanisms for air and water in a two-compartment model, labelled with their associated D-values. Vertical arrows indicate intercompartment transfer, and diagonal arrows indicate degradation losses from each compartment.

Isolating f_A from the second equation, we have:

$$f_A = \frac{D_{AW}^{Diff-Ov} f_W}{D_A^L}$$

Now substituting this expression for f_A into the expression for f_W we have:

$$f_W = \frac{E_W + D_{AW}^{T-A}\left[\frac{D_{AW}^{Diff-Ov} f_W}{D_A^L}\right]}{D_W^L}$$

or

$$f_W = \frac{E_W + \left[\frac{D_{AW}^{T-A} D_{AW}^{Diff-Ov} f_W}{D_A^L}\right]}{D_W^L}$$

Separating the two top terms as a sum of fractions:

$$f_W = \frac{E_W}{D_W^L} + \frac{D_{AW}^{T-A} D_{AW}^{Diff-Ov} f_W}{D_A^L D_W^L}$$

Bringing all the terms in f_W to the left-hand side:

$$f_W - \left[\frac{D_{AW}^{T-A} D_{AW}^{Diff-Ov}}{D_A^L D_W^L}\right] f_W = \frac{E_W}{D_W^L}$$

Factoring out f_W:

$$f_W \left[1 - \left(\frac{D_{AW}^{T-A} D_{AW}^{Diff-Ov}}{D_A^L D_W^L}\right)\right] = \frac{E_W}{D_W^L}$$

Isolating f_W:

$$f_W = \frac{E_W}{\left[1 - \left(\frac{D_{AW}^{T-A} D_{AW}^{Diff-Ov}}{D_A^L D_W^L}\right)\right] D_W^L}$$

Expanding the denominator and cancelling D_W^L in the second term:

$$f_W = \frac{E_W}{\left[D_W^L - \left(\frac{D_{AW}^{T-A} D_{AW}^{Diff-Ov}}{D_A^L}\right)\right]}$$

Bringing all terms in the denominator over a common denominator:

$$f_W = \frac{E_W}{\left[\frac{D_W^L D_A^L - D_{AW}^{T-A} D_{AW}^{Diff-Ov}}{D_A^L}\right]}$$

Bringing the denominator fraction's denominator up to the numerator:

$$f_W = \frac{E_W D_A^L}{D_W^L D_A^L - D_{AW}^{T-A} D_{AW}^{Diff-Ov}}$$

Back-substituting for D_W^L and expanding the term:

$$f_W = \frac{E_W D_A^L}{\left(D_{AW}^{Diff-Ov} D_A^L + D_W^{Deg} D_A^L\right) - D_{AW}^{T-A} D_{AW}^{Diff-Ov}}$$

Back-substituting for D_A^L in the first denominator term and D_{AW}^{T-A} in the third term to allow for cancellation of some terms, we have:

$$f_W = \frac{E_W D_A^L}{\left[D_{AW}^{Diff-Ov}\left(D_{AW}^{Diff-Ov} + D_A^{Dry} + D_A^{Wet} + D_A^{Rain} + D_A^{Deg}\right) + D_W^{Deg} D_A^L - D_{AW}^{Diff-Ov}\left(D_{AW}^{Diff-Ov} + D_A^{Dry} + D_A^{Wet} + D_A^{Rain}\right)\right]}$$

Cancelling common terms in the denominator, we arrive at:

$$\boxed{f_W = \frac{E_W D_A^L}{\left(D_{AW}^{Diff-Ov} D_A^{Deg} + D_W^{Deg} D_A^L\right)}}$$

With this solution for the water fugacity in hand, the air fugacity may be determined by way of either of the initial mass balances that relate it to the water fugacity. For example:

$$f_A = \frac{D_{AW}^{Diff-Ov} f_W}{D_A^L}$$

$$f_A = \frac{D_{AW}^{Diff-Ov}}{D_A^L}\left[\frac{E_W D_A^L}{\left(D_{AW}^{Diff-Ov} D_A^{Deg} + D_W^{Deg} D_A^L\right)}\right]$$

$$f_A = \frac{D_{AW}^{Diff-Ov} E_W}{\left(D_{AW}^{Diff-Ov} D_A^{Deg} + D_W^{Deg} D_A^L\right)}$$

$$\boxed{f_A = \frac{E_W}{\left[D_A^{Deg} + \left(\frac{D_W^{Deg} D_A^L}{D_{AW}^{Diff-Ov}}\right)\right]}}$$

Worked Example 4.8

A chemical is released at a rate of 3.00 *mol* h^{-1} into a water body that is in contact with an air body. The *D*-values for the various transport processes that connect the system are given below. Assume the system is at steady state and ignoring advection, use these *D*-values to determine the fugacity in the water and the air.

Transport Process	Symbol	D-value (*mol* Pa^{-1} h^{-1})
Air degradation	D_A^{Deg}	9.58×10^{-4}
Water degradation	D_W^{Deg}	3.03×10^{-4}
Air–water diffusion	$D_{AW}^{Diff-Ov}$	1.44×10^{-3}
Dry deposition	D_A^{Dry}	2.70×10^{-4}
Wet deposition	D_A^{Wet}	2.86×10^{-4}
Rain	D_A^{Rain}	1.14×10^{-6}

Begin by evaluating the total D-values for the loss processes in both water and air. The total of all loss processes for water is as follows:

$$D_W^L = \left(D_{AW}^{Diff-Ov} + D_W^{Deg}\right)$$

$$D_W^L = 1.44 \times 10^{-3} mol\, Pa^{-1} h^{-1} + 3.03 \times 10^{-4} mol\, Pa^{-1} h^{-1}$$

$$D_W^L = 1.74_3 \times 10^{-3} mol\, Pa^{-1} h^{-1}$$

For air, we have:

$$D_A^L = \left(D_{AW}^{Diff-Ov} + D_A^{Dry} + D_A^{Wet} + D_A^{Rain} + D_A^{Deg}\right)$$

$$D_A^L = \left(1.44 \times 10^{-3} + 2.70 \times 10^{-4} + 2.86 \times 10^{-4} + 1.14 \times 10^{-6} + 9.58 \times 10^{-4}\right) mol\, Pa^{-1} h^{-1}$$

$$D_A^L = 2.95_5 \times 10^{-3} mol\, Pa^{-1} h^{-1}$$

The total of all processes that transfer from air to water is:

$$D_{AW}^T = \left(D_{AW}^{Diff-Ov} + D_A^{Dry} + D_A^{Wet} + D_A^{Rain}\right)$$

$$D_{AW}^T = \left(1.44 \times 10^{-3} + 2.70 \times 10^{-4} + 2.86 \times 10^{-4} + 1.14 \times 10^{-6}\right) mol\, Pa^{-1} h^{-1}$$

$$D_{AW}^T = 1.99_7 \times 10^{-3} mol\, Pa^{-1} h^{-1}$$

With these in hand, we can solve for the water and air fugacities:

$$f_W = \frac{E_W D_A^L}{\left(D_{AW}^{Diff-Ov} D_A^{Deg} + D_W^{Deg} D_A^L\right)}$$

$$f_W = \frac{3.00\ mol\ h^{-1} \times 2.95_5 \times 10^{-3} mol\, Pa^{-1} h^{-1}}{\left(\begin{array}{l} 1.44 \times 10^{-3} mol\, Pa^{-1} h^{-1} \times 9.58 \times 10^{-4} mol\, Pa^{-1} h^{-1} \\ +3.03 \times 10^{-4} mol\, Pa^{-1} h^{-1} \times 2.95_5 \times 10^{-3} mol\, Pa^{-1} h^{-1} \end{array}\right)}$$

$$f_W = 3.90 \times 10^3\ Pa$$

$$f_A = \frac{E_W}{\left[D_A^{Deg} + \left(\dfrac{D_W^{Deg} D_A^L}{D_{AW}^{Diff-Ov}}\right)\right]}$$

$$f_A = \frac{3.00\ mol\ h^{-1}}{\left[9.58 \times 10^{-4} mol\, Pa^{-1} h^{-1} + \left(\dfrac{3.03 \times 10^{-4} mol\, Pa^{-1} h^{-1} \times 2.95_5 \times 10^{-3} mol\, Pa^{-1} h^{-1}}{1.44 \times 10^{-3} mol\, Pa^{-1} h^{-1}}\right)\right]}$$

$$f_A = 1.90 \times 10^3\ Pa$$

As the number of compartments increases, the number of independent equations containing the corresponding fugacities increases as well, and so too does the mathematical complexity and manipulation necessary to solve the problem. The essence of the approach is the same, in that a system of "n" compartments at steady state will have associated "n" independent equations that are functions of "n" fugacities. From that point on, algebraic substitution is usually performed to eliminate unknowns until one fugacity is determined. All other fugacities follow directly once the first is determined. With the fugacities in hand, all process rates can be determined and compartment concentrations can be calculated.

As an example of a four-compartment problem, the Level III EQC model environment will be solved in detail, developing the derivation results that are published in Mackay et al. (1992). In this model, air, water, soil, and sediment interact through various mechanisms. Bidirectional transport of chemicals is incorporated between air–water, air–soil and water–sediment pairs, as is unidirectional transport from soil to water. Degradation at unique rates is included in all four compartments, as are advective inputs for air, water, and sediment.

Symbolizing the D-value for the sum of all transport processes from compartment "i" into compartment "j" as D_{ij}^{T-i}, and from "j" to "i" as D_{ij}^{T-j}, the mass balances for the four compartments air, water, soil, and sediment are as follows, (as sum of inputs minus sum of outputs):

$$E_A + f_W D_{AW}^{T-W} + f_{Soil} D_{A-Soil}^{T-Soil} = f_A D_A^L$$
$$E_W + f_A D_{AW}^{T-A} + f_{Soil} D_{Soil-W}^{T-Soil} + f_{Sed} D_{Sed-W}^{T-Sed} = f_W D_W^L$$
$$E_{Soil} + f_A D_{A-Soil}^{T-A} = f_{Soil} D_{Soil}^L$$
$$E_{Sed} + f_W D_{Sed-W}^{T-W} = f_{Sed} D_{Sed}^L$$

Here D_i^L is the sum of all D-values for loss mechanisms for compartment "i". Specifically, these are:

$$D_A^L = D_A^{Deg} + D_A^{Adv} + D_{AW}^{T-A} + D_{A-Soil}^{T-A}$$
$$D_W^L = D_W^{Deg} + D_W^{Adv} + D_{AW}^{T-W} + D_{Sed-W}^{T-W}$$
$$D_{Soil}^L = D_{Soil}^{Deg} + D_{A-Soil}^{T-Soil} + D_{Soil-W}^{T-Soil}$$
$$D_{Sed}^L = D_{Sed}^{Deg} + D_{Sed}^{Adv} + D_{Sed-W}^{T-Sed}$$

We will seek to obtain an expression for the water fugacity f_W and express all the other fugacities in terms of f_w. Starting with the easier equations, we can define f_{Sed} in terms of f_W from the sediment mass balance:

$$E_{Sed} + f_W D_{Sed-W}^{T-W} = f_{Sed} D_{Sed}^L$$

$$f_{Sed} = \frac{E_{Sed} + f_W D_{Sed-W}^{T-W}}{D_{Sed}^L}$$

Similarly, we can isolate f_{Soil} in terms of f_A from the soil mass balance expression:

$$E_{Soil} + f_A D_{A-Soil}^{T-A} = f_{Soil} D_{Soil}^L$$

$$f_{Soil} = \frac{E_{Soil} + f_A D_{A-Soil}^{T-A}}{D_{Soil}^L}$$

From the air mass balance, we can develop an expression for f_A in terms of f_W, by first incorporating the expression for f_{Soil} from above:

$$E_A + f_W D_{AW}^{T-W} + f_{Soil} D_{A-Soil}^{T-Soil} = f_A D_A^L$$

$$E_A + f_W D_{AW}^{T-W} + \left(\frac{E_{Soil} + f_A D_{A-Soil}^{T-A}}{D_{Soil}^L} \right) D_{A-Soil}^{T-Soil} = f_A D_A^L$$

Expanding to "free up" the f_A term:

$$E_A + f_W D_{AW}^{T-W} + \left(\frac{E_{Soil} D_{A-Soil}^{T-Soil}}{D_{Soil}^L} \right) + f_A \left(\frac{D_{A-Soil}^{T-A} D_{A-Soil}^{T-Soil}}{D_{Soil}^L} \right) = f_A D_A^L$$

Bringing all terms with f_A to one side:

$$f_A D_A^L - f_A\left(\frac{D_{A-Soil}^{T-A} D_{A-Soil}^{T-Soil}}{D_{Soil}^L}\right) = E_A + f_W D_{AW}^{T-W} + \left(\frac{E_{Soil} D_{A-Soil}^{T-Soil}}{D_{Soil}^L}\right)$$

Factoring f_A:

$$f_A\left[D_A^L - \left(\frac{D_{A-Soil}^{T-A} D_{A-Soil}^{T-Soil}}{D_{Soil}^L}\right)\right] = E_A + f_W D_{AW}^{T-W} + \left(\frac{E_{Soil} D_{A-Soil}^{T-Soil}}{D_{Soil}^L}\right)$$

Isolating f_A:

$$f_A = \frac{E_A + f_W D_{AW}^{T-W} + \left(\frac{E_{Soil} D_{A-Soil}^{T-Soil}}{D_{Soil}^L}\right)}{\left[D_A^L - \left(\frac{D_{A-Soil}^{T-A} D_{A-Soil}^{T-Soil}}{D_{Soil}^L}\right)\right]}$$

Our aim is to have all fugacities expressed in terms of f_W, so we need to substitute this f_A expression into f_{Soil} to get:

$$f_{Soil} = \frac{E_{Soil} + f_A D_{A-Soil}^{T-A}}{D_{Soil}^L}$$

$$f_{Soil} = \left(\frac{E_{Soil}}{D_{Soil}^L}\right) + \left(\frac{E_A + f_W D_{AW}^{T-W} + \left(\frac{E_{Soil} D_{A-Soil}^{T-Soil}}{D_{Soil}^L}\right)}{\left[D_A^L - \left(\frac{D_{A-Soil}^{T-A} D_{A-Soil}^{T-Soil}}{D_{Soil}^L}\right)\right]}\right)\left(\frac{D_{A-Soil}^{T-A}}{D_{Soil}^L}\right)$$

Finally, substituting for f_A, f_{Soil} and f_{Sed} in the water mass balance with these expressions in terms of f_W, we have:

$$E_W + f_A D_{AW}^{T-A} + f_{Soil} D_{Soil-W}^{T-Soil} + f_{Sed} D_{Sed-W}^{T-Sed} = f_W D_W^L$$

$$E_W + \left(\frac{E_A + f_W D_{AW}^{T-W} + \left(\frac{E_{Soil} D_{A-Soil}^{T-Soil}}{D_{Soil}^L}\right)}{\left[D_A^L - \left(\frac{D_{A-Soil}^{T-A} D_{A-Soil}^{T-Soil}}{D_{Soil}^L}\right)\right]}\right) D_{AW}^{T-A}$$

$$+ \left(\left(\frac{E_{Soil}}{D_{Soil}^L}\right) + \left(\frac{E_A + f_W D_{AW}^{T-W} + \left(\frac{E_{Soil} D_{A-Soil}^{T-Soil}}{D_{Soil}^L}\right)}{\left[D_A^L - \left(\frac{D_{A-Soil}^{T-A} D_{A-Soil}^{T-Soil}}{D_{Soil}^L}\right)\right]}\right)\left(\frac{D_{A-Soil}^{T-A}}{D_{Soil}^L}\right)\right) D_{Soil-W}^{T-Soil}$$

$$+ \left(\frac{E_{Sed} + f_W D_{Sed-W}^{T-W}}{D_{Sed}^L}\right) D_{Sed-W}^{T-Sed} = f_W D_W^L$$

This expression could benefit from some cleaning up before we isolate f_W! First, multiplying through by the D-values outside the parentheses:

$$E_W + \left(\frac{E_A D_{AW}^{T-A} + f_W D_{AW}^{T-W} D_{AW}^{T-A} + \left(\frac{E_{Soil} D_{A-Soil}^{T-Soil} D_{AW}^{T-A}}{D_{Soil}^{L}} \right)}{\left[D_A^L - \left(\frac{D_{A-Soil}^{T-A} D_{A-Soil}^{T-Soil}}{D_{Soil}^{L}} \right) \right]} \right)$$

$$+ \left(\left(\frac{E_{Soil} D_{Soil-W}^{T-Soil}}{D_{Soil}^{L}} \right) + \left(\frac{E_A D_{Soil-W}^{T-Soil} + f_W D_{AW}^{T-W} D_{Soil-W}^{T-Soil} + \left(\frac{E_{Soil} D_{A-Soil}^{T-Soil} D_{Soil-W}^{T-Soil}}{D_{Soil}^{L}} \right)}{\left[D_A^L - \left(\frac{D_{A-Soil}^{T-A} D_{A-Soil}^{T-Soil}}{D_{Soil}^{L}} \right) \right]} \right) \left(\frac{D_{A-Soil}^{T-A}}{D_{Soil}^{L}} \right) \right)$$

$$+ \left(\frac{E_{Sed} D_{Sed-W}^{T-Sed} + f_W D_{Sed-W}^{T-W} D_{Sed-W}^{T-Sed}}{D_{Sed}^{L}} \right) = f_W D_W^L$$

Further expanding to eliminate the $\left(\frac{D_{A-Soil}^{T-A}}{D_{Soil}^{L}} \right)$ term:

$$E_W + \left(\frac{E_A D_{AW}^{T-A} + f_W D_{AW}^{T-W} D_{AW}^{T-A} + \left(\frac{E_{Soil} D_{A-Soil}^{T-Soil} D_{AW}^{T-A}}{D_{Soil}^{L}} \right)}{\left[D_A^L - \left(\frac{D_{A-Soil}^{T-A} D_{A-Soil}^{T-Soil}}{D_{Soil}^{L}} \right) \right]} \right)$$

$$+ \left(\left(\frac{E_{Soil} D_{Soil-W}^{T-Soil}}{D_{Soil}^{L}} \right) + \left(\frac{E_A D_{Soil-W}^{T-Soil} D_{A-Soil}^{T-A} + f_W D_{AW}^{T-W} D_{Soil-W}^{T-Soil} D_{A-Soil}^{T-A} + \left(\frac{E_{Soil} D_{A-Soil}^{T-Soil} D_{Soil-W}^{T-Soil} D_{A-Soil}^{T-A}}{D_{Soil}^{L}} \right)}{\left[D_A^L - \left(\frac{D_{A-Soil}^{T-A} D_{A-Soil}^{T-Soil}}{D_{Soil}^{L}} \right) \right] D_{Soil}^{L}} \right) \right)$$

$$+ \left(\frac{E_{Sed} D_{Sed-W}^{T-Sed} + f_W D_{Sed-W}^{T-W} D_{Sed-W}^{T-Sed}}{D_{Sed}^{L}} \right) = f_W D_W^L$$

Dividing by D_W^L and bringing f_W to the left-hand side:

$$f_W = \frac{E_W}{D_W^L} + \left(\frac{E_A D_{AW}^{T-A} + f_W D_{AW}^{T-W} D_{AW}^{T-A} + \left(\frac{E_{Soil} D_{A-Soil}^{T-Soil} D_{AW}^{T-A}}{D_{Soil}^{L}} \right)}{\left[D_W^L D_A^L - \left(\frac{D_W^L D_{A-Soil}^{T-A} D_{A-Soil}^{T-Soil}}{D_{Soil}^{L}} \right) \right]} \right)$$

$$+ \left(\left(\frac{E_{Soil} D_{Soil-W}^{T-Soil}}{D_{Soil}^{L} D_W^L} \right) + \left(\frac{E_A D_{Soil-W}^{T-Soil} D_{A-Soil}^{T-A} + f_W D_{AW}^{T-W} D_{Soil-W}^{T-Soil} D_{A-Soil}^{T-A} + \left(\frac{E_{Soil} D_{A-Soil}^{T-Soil} D_{Soil-W}^{T-Soil} D_{A-Soil}^{T-A}}{D_{Soil}^{L}} \right)}{\left[D_A^L - \left(\frac{D_{A-Soil}^{T-A} D_{A-Soil}^{T-Soil}}{D_{Soil}^{L}} \right) \right] D_{Soil}^{L} D_W^L} \right) \right)$$

$$+ \left(\frac{E_{Sed} D_{Sed-W}^{T-Sed} + f_W D_{Sed-W}^{T-W} D_{Sed-W}^{T-Sed}}{D_W^L D_{Sed}^{L}} \right)$$

At this point, the problem of determining f_W could be solved iteratively; however, an analytical solution is more useful and simpler to implement if one can obtain it. In this case, it is possible to obtain such a solution after some fairly messy algebra, which we now demonstrate.

Eliminating the denominator fractions:

$$f_W = \frac{E_W}{D_W^L} + \left(\frac{E_A D_{AW}^{T-A} D_{Soil}^L + f_W D_{AW}^{T-W} D_{AW}^{T-A} D_{Soil}^L + \left(\dfrac{E_{Soil} D_{A-Soil}^{T-Soil} D_{AW}^{T-A} D_{Soil}^L}{D_{Soil}^L} \right)}{\left[D_W^L D_A^L D_{Soil}^L - D_W^L D_{A-Soil}^{T-A} D_{A-Soil}^{T-Soil} \right]} \right)$$

$$+ \left(\left(\frac{E_{Soil} D_{Soil-W}^{T-Soil}}{D_W^L D_{Soil}^L} \right) + \left(\frac{E_A D_{Soil-W}^{T-Soil} D_{A-Soil}^{T-A} D_{Soil}^L + f_W D_{AW}^{T-W} D_{Soil-W}^{T-Soil} D_{A-Soil}^{T-A} D_{Soil}^L + \left(\dfrac{E_{Soil} D_{A-Soil}^{T-Soil} D_{Soil-W}^{T-Soil} D_{A-Soil}^{T-A} D_{Soil}^L}{D_{Soil}^L} \right)}{D_W^L D_{Soil}^L \left[D_A^L D_{Soil}^L - D_{A-Soil}^{T-A} D_{A-Soil}^{T-Soil} \right]} \right) \right)$$

$$+ \left(\frac{E_{Sed} D_{Sed-W}^{T-Sed} + f_W D_{Sed-W}^{T-W} D_{Sed-W}^{T-Sed}}{D_W^L D_{Sed}^L} \right)$$

Expanding to free up f_W terms:

$$f_W = \left(\frac{E_W}{D_W^L} \right) + \left(\frac{E_A D_{AW}^{T-A} D_{Soil}^L}{\left[D_W^L D_A^L D_{Soil}^L - D_W^L D_{A-Soil}^{T-A} D_{A-Soil}^{T-Soil} \right]} \right) + \left(\frac{f_W D_{AW}^{T-W} D_{AW}^{T-A} D_{Soil}^L}{\left[D_W^L D_A^L D_{Soil}^L - D_W^L D_{A-Soil}^{T-A} D_{A-Soil}^{T-Soil} \right]} \right)$$

$$+ \left(\frac{E_{Soil} D_{A-Soil}^{T-Soil} D_{AW}^{T-A} D_{Soil}^L}{D_{Soil}^L \left[D_W^L D_A^L D_{Soil}^L - D_W^L D_{A-Soil}^{T-A} D_{A-Soil}^{T-Soil} \right]} \right) + \left(\frac{E_{Soil} D_{Soil-W}^{T-Soil}}{D_W^L D_{Soil}^L} \right) + \left(\frac{E_A D_{Soil-W}^{T-Soil} D_{A-Soil}^{T-A} D_{Soil}^L}{D_W^L D_{Soil}^L \left[D_A^L D_{Soil}^L - D_{A-Soil}^{T-A} D_{A-Soil}^{T-Soil} \right]} \right)$$

$$+ \left(\frac{f_W D_{AW}^{T-W} D_{Soil-W}^{T-Soil} D_{A-Soil}^{T-A} D_{Soil}^L}{D_W^L D_{Soil}^L \left[D_A^L D_{Soil}^L - D_{A-Soil}^{T-A} D_{A-Soil}^{T-Soil} \right]} \right) + \left(\frac{E_{Soil} D_{A-Soil}^{T-Soil} D_{Soil-W}^{T-Soil} D_{A-Soil}^{T-A} D_{Soil}^L}{D_{Soil}^L D_W^L D_{Soil}^L \left[D_A^L D_{Soil}^L - D_{A-Soil}^{T-A} D_{A-Soil}^{T-Soil} \right]} \right)$$

$$+ \left(\frac{E_{Sed} D_{Sed-W}^{T-Sed}}{D_W^L D_{Sed}^L} \right) + \left(\frac{f_W D_{Sed-W}^{T-W} D_{Sed-W}^{T-Sed}}{D_W^L D_{Sed}^L} \right)$$

Cancelling terms and factoring the D-value for total water losses:

$$f_W = \left(\frac{E_W}{D_W^L} \right) + \left(\frac{E_A D_{AW}^{T-A} D_{Soil}^L}{D_W^L \left[D_A^L D_{Soil}^L - D_{A-Soil}^{T-A} D_{A-Soil}^{T-Soil} \right]} \right) + \left(\frac{f_W D_{AW}^{T-W} D_{AW}^{T-A} D_{Soil}^L}{D_W^L \left[D_A^L D_{Soil}^L - D_{A-Soil}^{T-A} D_{A-Soil}^{T-Soil} \right]} \right)$$

$$+ \left(\frac{E_{Soil} D_{A-Soil}^{T-Soil} D_{AW}^{T-A}}{D_W^L \left[D_A^L D_{Soil}^L - D_{A-Soil}^{T-A} D_{A-Soil}^{T-Soil} \right]} \right) + \left(\frac{E_{Soil} D_{Soil-W}^{T-Soil}}{D_W^L D_{Soil}^L} \right) + \left(\frac{E_A D_{Soil-W}^{T-Soil} D_{A-Soil}^{T-A}}{D_W^L \left[D_A^L D_{Soil}^L - D_{A-Soil}^{T-A} D_{A-Soil}^{T-Soil} \right]} \right)$$

$$+ \left(\frac{f_W D_{AW}^{T-W} D_{Soil-W}^{T-Soil} D_{A-Soil}^{T-A}}{D_W^L \left[D_A^L D_{Soil}^L - D_{A-Soil}^{T-A} D_{A-Soil}^{T-Soil} \right]} \right) + \left(\frac{E_{Soil} D_{A-Soil}^{T-Soil} D_{Soil-W}^{T-Soil} D_{A-Soil}^{T-A}}{D_W^L D_{Soil}^L \left[D_A^L D_{Soil}^L - D_{A-Soil}^{T-A} D_{A-Soil}^{T-Soil} \right]} \right)$$

$$+ \left(\frac{E_{Sed} D_{Sed-W}^{T-Sed}}{D_W^L D_{Sed}^L} \right) + \left(\frac{f_W D_{Sed-W}^{T-W} D_{Sed-W}^{T-Sed}}{D_W^L D_{Sed}^L} \right)$$

The term $D_W^L\left[D_A^L D_{Soil}^L - D_{A-Soil}^{T-A} D_{A-Soil}^{T-Soil}\right]$ appears many times and can be replaced by J_A:

$$f_W = \left(\frac{E_W}{D_W^L}\right) + \left(\frac{E_A D_{AW}^{T-A} D_{Soil}^L}{J_A}\right) + \left(\frac{f_W D_{AW}^{T-W} D_{AW}^{T-A} D_{Soil}^L}{J_A}\right) + \left(\frac{E_{Soil} D_{A-Soil}^{T-Soil} D_{AW}^{T-A}}{J_A}\right)$$
$$+\left(\frac{E_{Soil} D_{Soil-W}^{T-Soil}}{D_W^L D_{Soil}^L}\right) + \left(\frac{E_A D_{Soil-W}^{T-Soil} D_{A-Soil}^{T-A}}{J_A}\right) + \left(\frac{f_W D_{AW}^{T-W} D_{Soil-W}^{T-Soil} D_{A-Soil}^{T-A}}{J_A}\right)$$
$$+\left(\frac{E_{Soil} D_{A-Soil}^{T-Soil} D_{Soil-W}^{T-Soil} D_{A-Soil}^{T-A}}{D_{Soil}^L J_A}\right) + \left(\frac{E_{Sed} D_{Sed-W}^{T-Sed}}{D_W^L D_{Sed}^L}\right) + \left(\frac{f_W D_{Sed-W}^{T-W} D_{Sed-W}^{T-Sed}}{D_W^L D_{Sed}^L}\right)$$

Bringing all terms with f_W over to the left-hand side:

$$f_W - \left(\frac{f_W D_{AW}^{T-W} D_{AW}^{T-A} D_{Soil}^L}{J_A}\right) - \left(\frac{f_W D_{AW}^{T-W} D_{Soil-W}^{T-Soil} D_{A-Soil}^{T-A}}{J_A}\right) - \left(\frac{f_W D_{Sed-W}^{T-W} D_{Sed-W}^{T-Sed}}{D_W^L D_{Sed}^L}\right)$$
$$= \left(\frac{E_W}{D_W^L}\right) + \left(\frac{E_A D_{AW}^{T-A} D_{Soil}^L}{J_A}\right) + \left(\frac{E_{Soil} D_{A-Soil}^{T-Soil} D_{AW}^{T-A}}{J_A}\right)$$
$$+\left(\frac{E_{Soil} D_{Soil-W}^{T-Soil}}{D_W^L D_{Soil}^L}\right) + \left(\frac{E_A D_{Soil-W}^{T-Soil} D_{A-Soil}^{T-A}}{J_A}\right) + \left(\frac{E_{Soil} D_{A-Soil}^{T-Soil} D_{Soil-W}^{T-Soil} D_{A-Soil}^{T-A}}{D_{Soil}^L J_A}\right)$$
$$+\left(\frac{E_{Sed} D_{Sed-W}^{T-Sed}}{D_W^L D_{Sed}^L}\right)$$

Factoring out f_W:

$$f_W\left[1 - \left(\frac{D_{AW}^{T-W} D_{AW}^{T-A} D_{Soil}^L}{J_A}\right) - \left(\frac{D_{AW}^{T-W} D_{Soil-W}^{T-Soil} D_{A-Soil}^{T-A}}{J_A}\right) - \left(\frac{D_{Sed-W}^{T-W} D_{Sed-W}^{T-Sed}}{D_W^L D_{Sed}^L}\right)\right]$$
$$= \left(\frac{E_W}{D_W^L}\right) + \left(\frac{E_A D_{AW}^{T-A} D_{Soil}^L}{J_A}\right) + \left(\frac{E_{Soil} D_{A-Soil}^{T-Soil} D_{AW}^{T-A}}{J_A}\right)$$
$$+\left(\frac{E_{Soil} D_{Soil-W}^{T-Soil}}{D_W^L D_{Soil}^L}\right) + \left(\frac{E_A D_{Soil-W}^{T-Soil} D_{A-Soil}^{T-A}}{J_A}\right) + \left(\frac{E_{Soil} D_{A-Soil}^{T-Soil} D_{Soil-W}^{T-Soil} D_{A-Soil}^{T-A}}{D_{Soil}^L J_A}\right)$$
$$+\left(\frac{E_{Sed} D_{Sed-W}^{T-Sed}}{D_W^L D_{Sed}^L}\right)$$

Solving for f_W:

$$f_W = \frac{\left[\begin{array}{l}\left(\frac{E_W}{D_W^L}\right) + \left(\frac{E_A D_{AW}^{T-A} D_{Soil}^L}{J_A}\right) + \left(\frac{E_{Soil} D_{A-Soil}^{T-Soil} D_{AW}^{T-A}}{J_A}\right) \\ +\left(\frac{E_{Soil} D_{Soil-W}^{T-Soil}}{D_W^L D_{Soil}^L}\right) + \left(\frac{E_A D_{Soil-W}^{T-Soil} D_{A-Soil}^{T-A}}{J_A}\right) + \left(\frac{E_{Soil} D_{A-Soil}^{T-Soil} D_{Soil-W}^{T-Soil} D_{A-Soil}^{T-A}}{D_{Soil}^L J_A}\right) \\ +\left(\frac{E_{Sed} D_{Sed-W}^{T-Sed}}{D_W^L D_{Sed}^L}\right)\end{array}\right]}{\left[1 - \left(\frac{D_{AW}^{T-W} D_{AW}^{T-A} D_{Soil}^L}{J_A}\right) - \left(\frac{D_{AW}^{T-W} D_{Soil-W}^{T-Soil} D_{A-Soil}^{T-A}}{J_A}\right) - \left(\frac{D_{Sed-W}^{T-W} D_{Sed-W}^{T-Sed}}{D_W^L D_{Sed}^L}\right)\right]}$$

Back-substituting our expression for J_A, we have:

$$f_W = \frac{\left[\begin{array}{l} \left(\dfrac{E_W}{D^L_W}\right) + \left(\dfrac{E_A D^{T-A}_{AW} D^L_{Soil}}{D^L_W \left[D^L_A D^L_{Soil} - D^{T-A}_{A-Soil} D^{T-Soil}_{A-Soil}\right]}\right) + \left(\dfrac{E_{Soil} D^{T-Soil}_{A-Soil} D^{T-A}_{AW}}{D^L_W \left[D^L_A D^L_{Soil} - D^{T-A}_{A-Soil} D^{T-Soil}_{A-Soil}\right]}\right) \\ + \left(\dfrac{E_{Soil} D^{T-Soil}_{Soil-W}}{D^L_W D^L_{Soil}}\right) + \left(\dfrac{E_A D^{T-Soil}_{Soil-W} D^{T-A}_{A-Soil}}{D^L_W \left[D^L_A D^L_{Soil} - D^{T-A}_{A-Soil} D^{T-Soil}_{A-Soil}\right]}\right) \\ + \left(\dfrac{E_{Soil} D^{T-Soil}_{A-Soil} D^{T-Soil}_{Soil-W} D^{T-A}_{A-Soil}}{D^L_{Soil} D^L_W \left[D^L_A D^L_{Soil} - D^{T-A}_{A-Soil} D^{T-Soil}_{A-Soil}\right]}\right) + \left(\dfrac{E_{Sed} D^{T-Sed}_{Sed-W}}{D^L_W D^L_{Sed}}\right) \end{array} \right]}{\left[1 - \left(\dfrac{D^{T-W}_{AW} D^{T-A}_{AW} D^L_{Soil}}{D^L_W \left[D^L_A D^L_{Soil} - D^{T-A}_{A-Soil} D^{T-Soil}_{A-Soil}\right]}\right) - \left(\dfrac{D^{T-W}_{AW} D^{T-Soil}_{Soil-W} D^{T-A}_{A-Soil}}{D^L_W \left[D^L_A D^L_{Soil} - D^{T-A}_{A-Soil} D^{T-Soil}_{A-Soil}\right]}\right) - \left(\dfrac{D^{T-W}_{Sed-W} D^{T-Sed}_{Sed-W}}{D^L_W D^L_{Sed}}\right) \right]}$$

This expression can be simplified somewhat by factoring and multiplying numerator and denominator by D^L_W, resulting in:

$$f_W = \frac{\left[\begin{array}{l} E_W + \left(\dfrac{\left(E_A D^L_{Soil} + E_{Soil} D^{T-Soil}_{A-Soil}\right) D^{T-A}_{AW}}{\left[D^L_A D^L_{Soil} - D^{T-A}_{A-Soil} D^{T-Soil}_{A-Soil}\right]}\right) + \left(\dfrac{E_{Soil} D^{T-Soil}_{Soil-W}}{D^L_{Soil}}\right) \\ + \left(E_A + \dfrac{E_{Soil} D^{T-Soil}_{A-Soil}}{D^L_{Soil}}\right)\left(\dfrac{D^{T-Soil}_{Soil-W} D^{T-A}_{A-Soil}}{\left[D^L_A D^L_{Soil} - D^{T-A}_{A-Soil} D^{T-Soil}_{A-Soil}\right]}\right) + \left(\dfrac{E_{Sed} D^{T}_{Sed-W}}{D^L_{Sed}}\right) \end{array} \right]}{\left[D^L_W - \left(\dfrac{D^{T-W}_{AW} D^{T-A}_{AW} D^L_{Soil}}{\left[D^L_A D^L_{Soil} - D^{T-A}_{A-Soil} D^{T-Soil}_{A-Soil}\right]}\right) - \left(\dfrac{D^{T-W}_{AW} D^{T-Soil}_{Soil-W} D^{T-A}_{A-Soil}}{\left[D^L_A D^L_{Soil} - D^{T-A}_{A-Soil} D^{T-Soil}_{A-Soil}\right]}\right) - \left(\dfrac{D^{T-W}_{Sed-W} D^{T-Sed}_{Sed-W}}{D^L_{Sed}}\right) \right]}$$

To obtain the solution in the form published by Mackay et al. (1992) which uses simplifying substitutions for certain terms, we must do a few more algebraic steps. First the numerator terms are reordered to bring together the two terms containing E_A, which leads to this expression:

Bringing the terms in E_A together, we find:

$$f_W = \frac{\left[\begin{array}{l} E_W + \left(\dfrac{\left(E_A D^L_{Soil} + E_{Soil} D^{T-Soil}_{A-Soil}\right) D^{T-A}_{AW}}{\left[D^L_A D^L_{Soil} - D^{T-A}_{A-Soil} D^{T-Soil}_{A-Soil}\right]}\right) + \left(E_A + \dfrac{E_{Soil} D^{T-Soil}_{A-Soil}}{D^L_{Soil}}\right)\left(\dfrac{D^{T-Soil}_{Soil-W} D^{T-A}_{A-Soil}}{\left[D^L_A D^L_{Soil} - D^{T-A}_{A-Soil} D^{T-Soil}_{A-Soil}\right]}\right) \\ + \left(\dfrac{E_{Soil} D^{T-Soil}_{Soil-W}}{D^L_{Soil}}\right) + \left(\dfrac{E_{Sed} D^{T}_{Sed-W}}{D^L_{Sed}}\right) \end{array} \right]}{\left[D^L_W - \left(\dfrac{D^{T-W}_{AW} D^{T-A}_{AW} D^L_{Soil}}{\left[D^L_A D^L_{Soil} - D^{T-A}_{A-Soil} D^{T-Soil}_{A-Soil}\right]}\right) - \left(\dfrac{D^{T-W}_{AW} D^{T-Soil}_{Soil-W} D^{T-A}_{A-Soil}}{\left[D^L_A D^L_{Soil} - D^{T-A}_{A-Soil} D^{T-Soil}_{A-Soil}\right]}\right) - \left(\dfrac{D^{T-W}_{Sed-W} D^{T-Sed}_{Sed-W}}{D^L_{Sed}}\right) \right]}$$

Expanding the third numerator term we obtain:

$$f_W = \frac{\left[\begin{array}{l} E_W + \left(\dfrac{\left(E_A D_{Soil}^L + E_{Soil} D_{A-Soil}^{T-Soil}\right) D_{AW}^{T-A}}{\left[D_A^L D_{Soil}^L - D_{A-Soil}^{T-A} D_{A-Soil}^{T-Soil}\right]}\right) + \left(\dfrac{\left(E_A D_{Soil}^L + E_{Soil} D_{A-Soil}^{T-Soil}\right) D_{Soil-W}^{T-Soil} D_{A-Soil}^{T-A}}{D_{Soil}^L \left[D_A^L D_{Soil}^L - D_{A-Soil}^{T-A} D_{A-Soil}^{T-Soil}\right]}\right) \\ + \left(\dfrac{E_{Soil} D_{Soil-W}^{T-Soil}}{D_{Soil}^L}\right) + \left(\dfrac{E_{Sed} D_{Sed-W}^{T}}{D_{Sed}^L}\right) \end{array}\right]}{\left[D_W^L - \left(\dfrac{D_{AW}^{T-W} D_{AW}^{T-A} D_{Soil}^L}{\left[D_A^L D_{Soil}^L - D_{A-Soil}^{T-A} D_{A-Soil}^{T-Soil}\right]}\right) - \left(\dfrac{D_{AW}^{T-W} D_{Soil-W}^{T-Soil} D_{A-Soil}^{T-A}}{\left[D_A^L D_{Soil}^L - D_{A-Soil}^{T-A} D_{A-Soil}^{T-Soil}\right]}\right) - \left(\dfrac{D_{Sed-W}^{T-W} D_{Sed-W}^{T-Sed}}{D_{Sed}^L}\right)\right]}$$

Putting the second and third numerator terms over a common denominator and factoring $E_A D_{Soil}^L + E_{Soil} D_{A-Soil}^{T-Soil}$, we arrive at:

$$f_W = \frac{\left[E_W + \left(E_A D_{Soil}^L + E_{Soil} D_{A-Soil}^{T-Soil}\right)\left(\dfrac{D_{AW}^{T-A} + \left(\dfrac{D_{Soil-W}^{T-Soil} D_{A-Soil}^{T-A}}{D_{Soil}^L}\right)}{\left[D_A^L D_{Soil}^L - D_{A-Soil}^{T-A} D_{A-Soil}^{T-Soil}\right]}\right) + \left(\dfrac{E_{Soil} D_{Soil-W}^{T-Soil}}{D_{Soil}^L}\right) + \left(\dfrac{E_{Sed} D_{Sed-W}^{T}}{D_{Sed}^L}\right)\right]}{\left[D_W^L - \left(\dfrac{D_{AW}^{T-W} D_{AW}^{T-A} D_{Soil}^L}{\left[D_A^L D_{Soil}^L - D_{A-Soil}^{T-A} D_{A-Soil}^{T-Soil}\right]}\right) - \left(\dfrac{D_{AW}^{T-W} D_{Soil-W}^{T-Soil} D_{A-Soil}^{T-A}}{\left[D_A^L D_{Soil}^L - D_{A-Soil}^{T-A} D_{A-Soil}^{T-Soil}\right]}\right) - \left(\dfrac{D_{Sed-W}^{T-W} D_{Sed-W}^{T-Sed}}{D_{Sed}^L}\right)\right]}$$

Factoring the second and third denominator terms yields:

$$f_W = \frac{\left[E_W + \left(\dfrac{\left(E_A D_{Soil}^L + E_{Soil} D_{A-Soil}^{T-Soil}\right)\left(D_{AW}^{T-A} + \dfrac{D_{Soil-W}^{T-Soil} D_{A-Soil}^{T-A}}{D_{Soil}^L}\right)}{\left[D_A^L D_{Soil}^L - D_{A-Soil}^{T-Soil} D_{A-Soil}^{T-A}\right]}\right) + \left(\dfrac{E_{Soil} D_{Soil-W}^{T-Soil}}{D_{Soil}^L}\right) + \left(\dfrac{E_{Sed} D_{Sed-W}^{T-Sed}}{D_{Sed}^L}\right)\right]}{\left[D_W^L - \left(\dfrac{\left(D_{AW}^{T-W} D_{Soil}^L\right)\left(D_{AW}^{T-A} + \dfrac{D_{Soil-W}^{T-Soil} D_{A-Soil}^{T-A}}{D_{Soil}^L}\right)}{\left[D_A^L D_{Soil}^L - D_{A-Soil}^{T-Soil} D_{A-Soil}^{T-A}\right]}\right) - \left(\dfrac{D_{Sed-W}^{T-W} D_{Sed-W}^{T-Sed}}{D_{Sed}^L}\right)\right]}$$

To obtain the expression in the penultimate form, we must factor out unity in the form of $\left(\frac{D_A^L D_{Soil}^L}{D_A^L D_{Soil}^L}\right)$ from the top and bottom of the second terms of both the numerator and denominator to arrive at:

$$f_W = \frac{\left[E_W + \left(\dfrac{D_A^L D_{Soil}^L \left(\dfrac{E_A}{D_A^L} + \dfrac{E_{Soil} D_{A-Soil}^{T-Soil}}{D_{Soil}^L D_A^L}\right)\left(D_{AW}^{T-A} + \dfrac{D_{Soil-W}^{T-Soil} D_{A-Soil}^{T-A}}{D_{Soil}^L}\right)}{D_A^L D_{Soil}^L \left(1 - \dfrac{D_{A-Soil}^{T-Soil} D_{A-Soil}^{T-A}}{D_A^L D_{Soil}^L}\right)}\right) + \left(\dfrac{E_{Soil} D_{Soil-W}^{T-Soil}}{D_{Soil}^L}\right) + \left(\dfrac{E_{Sed} D_{Sed-W}^{T-Sed}}{D_{Sed}^L}\right)\right]}{\left[D_W^L - \left(\dfrac{D_A^L D_{Soil}^L \left(\dfrac{D_{AW}^{T-W}}{D_A^L}\right)\left(D_{AW}^{T-A} + \dfrac{D_{Soil-W}^{T-Soil} D_{A-Soil}^{T-A}}{D_{Soil}^L}\right)}{D_A^L D_{Soil}^L \left(1 - \dfrac{D_{A-Soil}^{T-Soil} D_{A-Soil}^{T-A}}{D_A^L D_{Soil}^L}\right)}\right) - \left(\dfrac{D_{Sed-W}^{T-W} D_{Sed-W}^{T-Sed}}{D_{Sed}^L}\right)\right]}$$

With cancellation of terms, this yields the following expression:

$$f_W = \frac{\left[E_W + \left(\frac{\left(\frac{E_A}{D_A^L} + \frac{E_{Soil} D_{A-Soil}^{T-Soil}}{D_{Soil}^L D_A^L} \right)\left(D_{AW}^{T-A} + \frac{D_{Soil-W}^{T-Soil} D_{A-Soil}^{T-A}}{D_{Soil}^L} \right)}{\left(1 - \frac{D_{A-Soil}^{T-Soil} D_{A-Soil}^{T-A}}{D_A^L D_{Soil}^L} \right)} \right) + \left(\frac{E_{Soil} D_{Soil-W}^{T-Soil}}{D_{Soil}^L} \right) + \left(\frac{E_{Sed} D_{Sed-W}^{T-Sed}}{D_{Sed}^L} \right) \right]}{\left[D_W^L - \left(\frac{\left(\frac{D_{AW}^{T-W}}{D_A^L} \right)\left(D_{AW}^{T-A} + \frac{D_{Soil-W}^{T-Soil} D_{A-Soil}^{T-A}}{D_{Soil}^L} \right)}{\left(1 - \frac{D_{A-Soil}^{T-Soil} D_{A-Soil}^{T-A}}{D_A^L D_{Soil}^L} \right)} \right) - \left(\frac{D_{Sed-W}^{T-W} D_{Sed-W}^{T-Sed}}{D_{Sed}^L} \right) \right]}$$

The final expression for f_W, in the format published by Mackay et al. (1992), gives the water and air fugacities in a more compact form by equating certain recurring terms in this expression with the symbols J_1 to J_4. These are defined as follows:

$$J_1 = \left(\frac{E_A}{D_A^L} \right) + \left(\frac{E_{Soil} D_{A-Soil}^{T-Soil}}{D_{Soil}^L D_A^L} \right)$$

$$J_2 = \left(\frac{D_{AW}^{T-W}}{D_A^L} \right)$$

$$J_3 = 1 - \left(\frac{D_{A-Soil}^{T-Soil} D_{A-Soil}^{T-A}}{D_A^L D_{Soil}^L} \right)$$

$$J_4 = D_{AW}^{T-A} + \left(\frac{D_{Soil-W}^{T-Soil} D_{A-Soil}^{T-A}}{D_{Soil}^L} \right)$$

Substitution of these into the expression for the water compartment fugacity leads to the final expressions given in Mackay et al. (1992):

$$f_W = \frac{\left[E_W + (J_1 J_4 / J_3) + \left(E_{Soil} D_{Soil-W}^{T-Soil} / D_{Soil}^L \right) + \left(E_{Sed} D_{Sed-W}^{T-Sed} / D_{Sed}^L \right) \right]}{\left[D_W^L - (J_2 J_4 / J_3) - \left(D_{Sed-W}^{T-W} D_{Sed-W}^{T-Sed} / D_{Sed}^L \right) \right]}$$

Once f_W is determined, the solutions to the other fugacities follow directly from the equations which define them above in terms of f_W:

$$f_A = \frac{E_A + f_W D_{AW}^{T-W} + \left(\frac{E_{Soil} D_{A-Soil}^{T-Soil}}{D_{Soil}^L} \right)}{\left[D_A^L - \left(\frac{D_{A-Soil}^{T-A} D_{A-Soil}^{T-Soil}}{D_{Soil}^L} \right) \right]}$$

Substituting for the same J values, we have also:

$$f_A = \frac{\left[J_1 + f_W J_2 \right]}{J_3}$$

For soil and sediment, we obtain directly:

$$f_{Soil} = \frac{E_{Soil} + f_A D_{A-Soil}^{T-A}}{D_{Soil}^{L}}$$

$$f_{Sed} = \frac{E_{Sed} + f_W D_{Sed-W}^{T-W}}{D_{Sed}^{L}}$$

Worked Example 4.9

Using the following tabulated *D*-values for the various transport processes for benzo[a]pyrene (molar mass 252.3 *g mol*$^{-1}$) in the EQC Level III environment, calculate the chemical's fugacities of each of the four compartments of air, water, soil, and sediment from a 1000 *kg h*$^{-1}$ release into (1) the air, (2) the water, and (3) the soil.

Transport Processes	Symbol	*D*-value (*mol Pa*$^{-1}$ *h*$^{-1}$)
Air advection	D_A^{Adv}	2.35×10^9
Water advection	D_W^{Adv}	7.44×10^9
Sediment advection	D_{Sed}^{Adv}	1.86×10^9
Air degradation	D_A^{Deg}	9.58×10^8
Water degradation	D_W^{Deg}	3.03×10^9
Soil degradation	D_{Soil}^{Deg}	1.70×10^{11}
Sediment degradation	D_{Sed}^{Deg}	1.17×10^9
Air to water	D_{AW}^{T-A}	6.24×10^8
Air to soil	D_{A-Soil}^{T-A}	5.45×10^9
Water to air	D_{AW}^{T-W}	2.05×10^7
Water to sediment	D_{Sed-W}^{T-W}	1.45×10^{10}
Soil to air	D_{A-Soil}^{T-Soil}	1.81×10^7
Soil to water	D_{Soil-W}^{T-Soil}	5.14×10^8
Sediment to water	D_{Sed-W}^{T-Sed}	1.88×10^9

We shall look at the first case of release into the air as a detailed example, and then present results for the other two cases in tabular form for comparison.

We will need the rate of chemical release in molar units, so first we must calculate the molar rate of release from the mass rate:

$$E_i = \frac{mass\,rate}{M} = \frac{1000.\,kg\,h^{-1}}{252.3\,g\,mol^{-1} \times 10^{-3}\,kg\,g^{-1}} = 3.963_5 \times 10^3\,mol\,h^{-1}$$

To calculate the fugacities, we need to know the "*J*" values, but to calculate the "*J*" values, we first need to know the *D*-values for the total loss from each of the four compartments:

$$D_A^L = D_A^{Deg} + D_A^{Adv} + D_{AW}^{T-A} + D_{A-Soil}^{T-A}$$

$$D_A^L = \left(9.58 \times 10^8 + 2.35 \times 10^9 + 6.24 \times 10^8 + 5.45 \times 10^9\right) = 9.38_2 \times 10^9\,mol\,Pa^{-1}h^{-1}$$

$$D_W^L = D_W^{Deg} + D_W^{Adv} + D_{AW}^{T-W} + D_{Sed-W}^{T-W}$$

$$D_W^L = 3.03 \times 10^9 + 7.44 \times 10^9 + 2.05 \times 10^7 + 1.45 \times 10^{10} = 2.49_9 \times 10^{10}\,mol\,Pa^{-1}h^{-1}$$

$$D_{Soil}^L = D_{Soil}^{Deg} + D_{A-Soil}^{T-Soil} + D_{Soil-W}^{T-Soil}$$

$$D_{Soil}^L = 1.70 \times 10^{11} + 1.81 \times 10^7 + 5.14 \times 10^8 = 1.70_5 \times 10^{11}\,mol\,Pa^{-1}h^{-1}$$

$$D_{Sed}^L = D_{Sed}^{Deg} + D_{Sed}^{Adv} + D_{Sed-W}^{T-Sed}$$

$$D_{Sed}^L = 1.17 \times 10^9 + 1.86 \times 10^9 + 1.88 \times 10^9 = 4.91_0 \times 10^9\,mol\,Pa^{-1}h^{-1}$$

Using the equations from Mackay et al. (1992), we have the following "J" values:

$$J_1 = \frac{E_A}{D_A^L} + \frac{E_{Soil} D_{A-Soil}^{T-Soil}}{D_{Soil}^L D_A^L}$$

$$J_1 = \frac{3.963_5 \times 10^3 mol\ h^{-1}}{9.38_2 \times 10^9 mol\ Pa^{-1}h^{-1}} + \frac{0\ mol\ h^{-1} \times 1.81 \times 10^7 mol\ Pa^{-1}h^{-1}}{1.70_5 \times 10^{11} mol\ Pa^{-1}h^{-1} \times 9.38_2 \times 10^9 mol\ Pa^{-1}h^{-1}} = 4.22_5 \times 10^{-7}\ Pa$$

$$J_2 = \frac{D_{AW}^{T-W}}{D_A^L}$$

$$J_2 = \frac{2.05 \times 10^7 mol\ Pa^{-1}h^{-1}}{9.38_2 \times 10^9 mol\ Pa^{-1}h^{-1}} = 2.18_5 \times 10^{-3}$$

$$J_3 = 1 - \frac{D_{A-Soil}^{T-Soil} D_{A-Soil}^{T-A}}{D_A^L D_{Soil}^L}$$

$$J_3 = 1 - \frac{1.81 \times 10^7 mol\ Pa^{-1}h^{-1} \times 5.45 \times 10^9 mol\ Pa^{-1}h^{-1}}{9.38_2 \times 10^9 mol\ Pa^{-1}h^{-1} \times 1.70_5 \times 10^{11} mol\ Pa^{-1}h^{-1}} = 9.99_9 \times 10^{-1}$$

$$J_4 = D_{AW}^{T-A} + \frac{D_{Soil-W}^{T-Soil} D_{A-Soil}^{T-A}}{D_{Soil}^L}$$

$$J_4 = 6.24 \times 10^8 mol\ Pa^{-1}h^{-1} + \frac{5.14 \times 10^8 mol\ Pa^{-1}h^{-1} \times 5.45 \times 10^9 mol\ Pa^{-1}h^{-1}}{1.70_5 \times 10^{11} mol\ Pa^{-1}h^{-1}} = 6.40_4 \times 10^8 mol\ Pa^{-1}h^{-1}$$

With the "J" values in hand, we can now calculate the various compartment fugacities. For readability, the units associated with the various values will not be included (but should always be checked!):

$$f_W = \frac{\left[E_W + \left(\frac{J_1 J_4}{J_3}\right) + \left(\frac{E_{Soil} D_{Soil-W}^{T-Soil}}{D_{Soil}^L}\right) + \left(\frac{E_{Sed} D_{Sed-W}^{T-Sed}}{D_{Sed}^L}\right)\right]}{\left[D_W^L - \left(\frac{J_2 J_4}{J_3}\right) - \left(\frac{D_{Sed-W}^{T-W} D_{Sed-W}^{T-Sed}}{D_{Sed}^L}\right)\right]}$$

The numerator in this expression for f_W is:

$$f_W(Num) = \left[\begin{array}{l} 0 + \left(\frac{4.22_5 \times 10^{-7} \times 6.40_4 \times 10^8}{9.99_9 \times 10^{-1}}\right) \\ + \left(\frac{0.0 \times 5.14 \times 10^8}{1.70_5 \times 10^{11}}\right) + \left(\frac{0 \times 1.88 \times 10^9}{4.91_0 \times 10^9}\right) \end{array}\right]$$

$$f_W(Num) = 2.70_6 \times 10^2$$

The denominator is:

$$f_W(Den) = \left[\begin{array}{l} 2.49_9 \times 10^{10} - \left(\frac{2.18_5 \times 10^{-3} \times 6.40_4 \times 10^8}{9.99_9 \times 10^{-1}}\right) \\ - \left(\frac{1.45 \times 10^{10} \times 1.88 \times 10^9}{4.91_0 \times 10^9}\right) \end{array}\right]$$

$$f_W(Den) = 1.94_4 \times 10^{10}$$

The fugacity in water is therefore:

$$f_W = \frac{f_W(Num)}{f_W(Den)} = \frac{2.70_6 \times 10^2}{1.94_4 \times 10^{10}} = 1.39_2 \times 10^{-8}\ Pa$$

The remaining fugacities are calculated directly from their corresponding expressions, f_A and f_{Sed} being dependent on the value of f_W calculated above, and f_{Soil} being dependent on the value of f_A.

$$f_A = \frac{[J_1 + f_W J_2]}{J_3}$$

$$f_A = \frac{[4.22_5 \times 10^{-7} + 1.39_2 \times 10^{-8} \times 2.18_5 \times 10^{-3}]}{9.99_9 \times 10^{-1}} = 4.22_5 \times 10^{-7}\ Pa$$

$$f_{Soil} = \frac{[E_{Soil} + f_A D^{T-A}_{A-Soil}]}{D^L_{Soil}}$$

$$f_{Soil} = \frac{[0 + 4.22_5 \times 10^{-7} \times 5.45 \times 10^9]}{1.70_5 \times 10^{11}} = 1.35_0 \times 10^{-8}\ Pa$$

$$f_{Sed} = \frac{[E_{Sed} + f_W D^{T-W}_{Sed-W}]}{D^L_{Sed}}$$

$$f_{Sed} = \frac{[0 + 1.39_2 \times 10^{-8} \times 1.45 \times 10^{10}]}{4.91_0 \times 10^9} = 4.11_1 \times 10^{-8}\ Pa$$

For all three cases, the results are tabulated below and are calculated in the same way:

	Emission to Air	Emission to Water	Emission to Soil
J_1	$4.22_5 \times 10^{-7}$	0	$4.48_4 \times 10^{-11}$
J_2	$2.18_5 \times 10^{-3}$	$2.18_5 \times 10^{-3}$	$2.18_5 \times 10^{-3}$
J_3	$9.99_9 \times 10^{-1}$	$9.99_9 \times 10^{-1}$	$9.99_9 \times 10^{-1}$
J_4	$6.40_4 \times 10^8$	$6.40_4 \times 10^8$	$6.40_4 \times 10^8$
f_W (N)	$2.70_6 \times 10^2$	$3.96_4 \times 10^3$	$1.19_8 \times 10^1$
f_W (D)	$1.94_4 \times 10^{10}$	$1.94_4 \times 10^{10}$	$1.94_4 \times 10^{10}$
f_W	$1.39_2 \times 10^{-8}$ Pa	$2.03_9 \times 10^{-7}$ Pa	$6.16_1 \times 10^{-10}$ Pa
f_A	$4.22_5 \times 10^{-7}$ Pa	$4.45_6 \times 10^{-10}$ Pa	$4.61_9 \times 10^{-11}$ Pa
f_{Soil}	$1.35_0 \times 10^{-8}$ Pa	$1.42_4 \times 10^{-11}$ Pa	$2.32_4 \times 10^{-8}$ Pa
f_{Sed}	$4.11_1 \times 10^{-8}$ Pa	$6.02_2 \times 10^{-7}$ Pa	$1.81_9 \times 10^{-9}$ Pa

It is clear that the fugacities are significantly influenced by the location of the chemical's release. For example, the sediment fugacity is much higher when the chemical is released to water, from which it experiences the greatest fugacity gradient, whereas it is lowest when the chemical is released to the soil.

With these fugacities in hand, the rates for all processes can be determined by the product $r = Df$, concentrations in each compartment can be determined by the products $C = Zf$, and amounts in each compartment can be determined by the products VZf. The reader can verify these results using the EQC program with benzo[a]pyrene and the EQC standard environment.

5 Basic Environmental Models
Putting It All Together

The different types of steady-state Mackay models (Level I, Level II, and Level III) are discussed and elucidated in this chapter. Default input values from the Canadian Environmental Modelling Centre (CEMC) models by these names are shown, and the computations involved are explained fully. The necessary fugacity capacities and D-values are developed as needed, in sequence, for one model chemical system. Equations for determining specific and bulk fugacity capacities, flow rates, and D-values for many processes such as advection, degradation, dry and wet deposition, sediment deposition and resuspension, and soil water and solids runoff are all considered. Total concentrations and mass or molar quantities are determined for each model.

5.1 LEVEL I MODELS

The simplest type of environmental model involves a closed system in which two or more compartments are at equilibrium. This is termed a "Level I" model in the Mackay nomenclature system. A Level I model is useful to get a sense of where a chemical "would want to go", if equilibrium were achieved in a model environment. As well, it requires relatively little data to run compared to higher-level models. Transport processes introduced at higher modelling levels will alter the distribution of the chemical in a quantitative sense, but, from a qualitative point of view, the Level I result gives a useful rough estimate of how a chemical will be distributed.

The strategy in a Level I modelling scenario is to first determine the fugacity capacities of all compartments, and then to use these, along with the compartment volumes and the amount of chemical in the system, to determine the prevailing system fugacity:

$$\boxed{f_{Sys} = \frac{m}{\sum_{i=1}^{n} Z_i V_i}}$$

Once the prevailing system fugacity has been determined, compartment concentrations, mole or mass fractions or percentages, and total amounts in moles or mass can easily be calculated.

As a working example, we will follow a complete Level I model calculation for the chemical benzo[a]pyrene. The Level I model encompasses seven compartments, or four media (air, water, soil, and sediment), two of which have subcompartments (aerosols and pure air in an air-based medium, and suspended solids, fish, and pure water in a water-based medium). For input data, we will use values from the updated standard EQC (Equilibrium Criterion) environment (Hughes et al. 2012).

We first set out to determine the various fugacity capacities of each compartment. The fugacity capacities of the soil, sediment, and suspended solids are modelled by assuming they are each equal to that of organic carbon multiplied by the mass fraction of organic carbon in the compartment, thereby ignoring all other non-organic content. The fish fugacity capacity is calculated in the manner introduced in Chapter 1, based on the lipid content.

The necessary input information for the calculation, as it appears in the Level I spreadsheet version available from the CEMC (https://www.trentu.ca/cemc/resources-and-models), is shown in Figure 5.1.

DOI: 10.1201/9781003657170-5

CHEMICAL PROPERTIES **Please complete all required fields below or Select a chemical from the database:**

*** indicates required input**

Note: insert dash (-) if property not inputted

*	Chemical Name	Benzo[a]pyrene
	CAS	
*	Molar Mass (g/mol)	252.316
*	Data Temperature (°C)	25
	Melting Point (°C)	175
	Vapor Pressure (Pa)	7.00E-07
*	Henry's Law Constant (Pa·m³/mol)	0.0463

Partition Ratios

	logK_{OW}	6.04
*	K_{OC}	3.838E+05
	-	
	-	
	-	
	-	
	-	
	-	

References and Notes

Seth method

Click **Add to DB** to add this chemical to the Chemical Database

FIGURE 5.1 Screenshot of the Chemical Properties screen from the Level I model available from the CEMC website, with required input values for benzo[a]pyrene.

The EQC environmental parameters, as updated by Hughes et al. (2012), are as shown below in Figure 5.2. Note that the EQC environment is a standard environment and none of the environmental parameters should be changed if the EQC environment is to be invoked.

The amount of chemical released into the system is $1.000 \times 10^5\ kg$. Since the system is closed and at equilibrium, the release location does not need to be defined. The calculation assumes that the chemical is fully distributed and equilibrated throughout the system, and the release location has no effect in this scenario. The input screen for the amount of chemical released is shown in Figure 5.3.

The air fugacity capacity is given by:

$$\boxed{Z_A = \frac{1}{RT}}$$

$$Z_A = \frac{1}{8.31446\,m^3\,Pa\,K^{-1}mol^{-1} \times (25.0 + 273.15)\,K} = 4.034_0 \times 10^{-4}\,mol\,Pa^{-1}\,m^{-3}$$

ENVIRONMENTAL PROPERTIES

Please complete all fields or select an environment from the database:

Environment Name	EQC- 2012-Hughes et al.

Fractions

Aerosols in Air	2.00E-11
Susp. Particles in Water	5.00E-06
Fish in Water	1.00E-06

Volume

Air (m³)	1.00E+14
Water (m³)	2.00E+11
Soil (m³)	1.80E+10
Sediment (m³)	5.00E+08

Lipid Fraction

Fish (g/g)	0.05

Organic Carbon

Suspended Particles (g/g)	0.2
Soil (g/g)	0.02
Sediment (g/g)	0.04

Density

Air (kg/m³)	1.2
Aerosol (kg/m³)	2000
Water (kg/m³)	1000
Suspended Particles (kg/m³)	1500
Fish (kg/m³)	1000
Soil (kg/m³)	2400
Sediment (kg/m³)	2400

FIGURE 5.2 Screenshot of the Environmental Properties screen from the Level I model, with input values for the updated standard EQC environment.

CHEMICAL QUANTITY

	Kilograms	Mols
Amount of Chemical	100000	396328.4

FIGURE 5.3 Screenshot of the Chemical Quantity screen from the Level I model, with the default input value for amount of benzo[a]pyrene used.

The water fugacity capacity can be determined from the Henry's Law constant for benzo[a]pyrene:

$$\boxed{Z_W = \frac{1}{H}}$$

$$Z_W = \frac{1}{4.63\times10^{-2}\,Pa\,m^3 mol^{-1}} = 21.6_0\,mol\,Pa^{-1}\,m^{-3}$$

The soil fugacity capacity is given by:

$$Z_{Soil} = m_{Soil}^{f-OC} \times 0.35\,L\ kg^{-1} \times K_{OW} \times Z_W \times \left(\frac{\rho_{Soil}\left(kg\,m^{-3}\right)}{1000\,L\,m^{-3}}\right)$$

$$Z_{Soil} = 0.020 \times 0.35\,L\ kg^{-1} \times 10^{6.04} \times 21.6_0\,mol\,Pa^{-1}\,m^{-3} \times \left(\frac{2400\,kg\,m^{-3}}{1000\,L\,m^{-3}}\right)$$

$$Z_{Soil} = 3.9_8 \times 10^5\,mol\,Pa^{-1}\,m^{-3}$$

The sediment fugacity capacity is:

$$Z_{Sed} = m_{Sed}^{f-OC} \times 0.35\,L\ kg^{-1} \times K_{OW} \times Z_W \times \left(\frac{\rho_{Sed}\left(kg\,m^{-3}\right)}{1000\,L\,m^{-3}}\right)$$

$$Z_{Sed} = 0.040 \times 0.35\,L\ kg^{-1} \times 10^{6.04} \times 21.6_0\,mol\,Pa^{-1}\,m^{-3} \times \left(\frac{2400\,kg\,m^{-3}}{1000\,L\,m^{-3}}\right)$$

$$Z_{Sed} = 7.9_6 \times 10^5\,mol\,Pa^{-1}\,m^{-3}$$

The aerosols fugacity capacity is estimated in the Level I model by the Mackay expression (Mackay et al. 1986) introduced in Chapter 2.

$$Z_Q = \left(\frac{6\times10^6 Pa}{P_A^{Sat}(Pa)}\right) Z_A$$

To use this expression, we first need to calculate the subcooled saturation vapour pressure of benzo[a]pyrene, since its melting point is above the modelling temperature. Using the expression for fugacity ratio introduced in Chapter 1, we have:

$$\ln F \simeq 6.79\left(1 - \frac{T_M}{T(K)}\right)$$

$$F = \exp\left(6.79\left(1 - \frac{(175+273.15)K}{(25.0+273.15)K}\right)\right) = 3.2_8 \times 10^{-2}$$

With the fugacity ratio in hand, we can now calculate the subcooled liquid vapour pressure:

$$P_A^{Sat-SCL} = \frac{P_A^{Sat}}{F} = \frac{7.00\times 10^{-7}\,Pa}{3.2_8\times 10^{-2}} = 2.1_3\times 10^{-5}\,Pa$$

Finally, the aerosol fugacity capacity is given by:

$$Z_Q = \left(\frac{6\times 10^6\,Pa}{2.1_3\times 10^{-5}\,Pa}\right)\times 4.034_0\times 10^{-4}\,mol\,Pa^{-1}\,m^{-3} = 1.1_4\times 10^8\,mol\,Pa^{-1}\,m^{-3}$$

The suspended solids fugacity capacity is given by:

$$Z_{SS} = m_{SS}^{f-OC}\times 0.35\,L\ kg^{-1}\times K_{OW}\times Z_W\times\left(\frac{\rho_{SS}\left(kg\,m^{-3}\right)}{1000\,L\,m^{-3}}\right)$$

$$Z_{SS} = 0.20\times 0.35\,L\ kg^{-1}\times 10^{6.04}\times 21.6_0\,mol\,Pa^{-1}\,m^{-3}\times\left(\frac{1500\,kg\,m^{-3}}{1000\,L\,m^{-3}}\right)$$

$$Z_{SS} = 2.4_9\times 10^6\,mol\,Pa^{-1}\,m^{-3}$$

The fugacity capacity of fish is given in terms of the fish–water partition ratio, which is given by:

$$K_{Fish-W} = v_{Fish}^{f-L}K_{OW}$$

Since $K_{Fish-W} = Z_{Fish}/Z_W$ the bulk fugacity capacity of the fish follows as:

$$Z_{Fish} = K_{Fish-W}Z_W = v_{Fish}^{f-L}K_{OW}Z_W$$

$$Z_{Fish} = 0.050\times 10^{6.04}\times 21.6_0\,mol\,Pa^{-1}\,m^{-3}$$

$$Z_{Fish} = 1.1_8\times 10^6\,mol\,Pa^{-1}\,m^{-3}$$

We can now solve for the system fugacity, according to the expression given above:

$$\boxed{f_{Sys} = \frac{m}{\sum_{i=1}^{n} Z_i V_i}}$$

The total moles of chemical in the system is:

$$m = \frac{emission\,mass}{M} = \frac{1.000\times 10^5\,kg}{252.316\,g\,mol^{-1}\times 10^{-3}\,kg\,g^{-1}} = 3.963_3\times 10^5\,mol$$

The various compartment volumes are calculated either as the product of the appropriate volume fraction and the volume of the carrier medium (see Figure 5.2 for data), or by difference based on the volume of the minor components and the bulk volume of the medium.

For air:

$$V_A^A = V_A^{Bulk} - V_A^Q$$

$$V_A^A = V_A^{Bulk} - \left(v_A^{f-Q}V_A^{Bulk}\right)$$

$$V_A^A = V_A^{Bulk}\left(1 - v_A^{f-Q}\right)$$

$$V_A^A = 1.00\times 10^{14}\,m^3\left(1 - 2.00\times 10^{-11}\right) = 1.00_0\times 10^{14}\,m^3\left(= V_A^{Bulk}\right)$$

For aerosols:

$$V_Q = v_A^{f-Q} V_A^{Bulk} = 2.00 \times 10^{-11} \times 1.00 \times 10^{14}\, m^3 = 2.00_0 \times 10^3\, m^3$$

For water:

$$V_W^W = V_W^{Bulk} - \left(V_W^{SS} + V_W^{Fish}\right)$$

$$V_W^W = V_W^{Bulk} - \left(v_W^{f-SS} V_W^{Bulk} + v_W^{f-Fish} V_W^{Bulk}\right)$$

$$V_W^W = V_W^{Bulk}\left(1 - v_W^{f-SS} - v_W^{f-Fish}\right)$$

$$V_W^W = 2.00 \times 10^{11}\, m^3 \left(1 - 5.00 \times 10^{-6} - 1.00 \times 10^{-6}\right)$$

$$V_W^W = 2.00_0 \times 10^{11}\, m^3 = \left(V_W^{Bulk}\right)$$

For suspended solids and fish:

$$V_{SS} = v_W^{f-SS} V_W^{Bulk} = 5.00 \times 10^{-6} \times 2.00 \times 10^{11}\, m^3 = 1.00_0 \times 10^6\, m^3$$

$$V_{Fish} = v_W^{f-Fish} V_W^{Bulk} = 1.00 \times 10^{-6} \times 2.00 \times 10^{11}\, m^3 = 2.00_0 \times 10^5\, m^3$$

The volume of soil and sediment are equal to their bulk medium volumes, since they do not contain any subcompartment components that are considered in this model:

$$V_{Soil} = 1.80 \times 10^{10}\, m^3$$

$$V_{Sed} = 5.00 \times 10^8\, m^3$$

Word of Warning! Since the volume fractions of the aerosols in air, as well as suspended solids and fish in water, are small, some models will simply assign the total bulk air and water volumes to the "air-in-air" and "water-in-water" compartments. While this yields the same result, it may lead to interpretive confusion if not anticipated.

The sum of ZV products is given by:

$$\sum_{i=1}^{n} Z_i V_i = Z_A V_A + Z_Q V_Q + Z_W V_W + Z_{SS} V_{SS} + Z_{Fish} V_{Fish} + Z_{Soil} V_{Soil} + Z_{Sed} V_{Sed}$$

$$\sum_{i=1}^{n} Z_i V_i = \left[\begin{array}{l} 4.034_0 \times 10^{-4}\, mol\, Pa^{-1}\, m^{-3} \times 1.00_0 \times 10^{14}\, m^3 + 1.1_4 \times 10^8\, mol\, Pa^{-1}\, m^{-3} \times 2.00_0 \times 10^3\, m^3 \\ +21.6_0\, mol\, Pa^{-1}\, m^{-3} \times 2.00_0 \times 10^{11}\, m^3 + 2.4_9 \times 10^6\, mol\, Pa^{-1}\, m^{-3} \times 1.00_0 \times 10^6\, m^3 \\ +1.1_8 \times 10^6\, mol\, Pa^{-1}\, m^{-3} \times 2.00_0 \times 10^5\, m^3 + 3.9_8 \times 10^5\, mol\, Pa^{-1}\, m^{-3} \times 1.80 \times 10^{10}\, m^3 \\ +7.9_6 \times 10^5\, mol\, Pa^{-1}\, m^{-3} \times 5.00 \times 10^8\, m^3 \end{array}\right]$$

$$\sum_{i=1}^{n} Z_i V_i = 7.5_7 \times 10^{15}\, mol\, Pa^{-1}$$

The system fugacity is therefore:

$$\boxed{f_{Sys} = \frac{m}{\sum_{i=1}^{n} Z_i V_i}}$$

$$f_{Sys} = \frac{3.963_3 \times 10^5\, mol}{7.5_7 \times 10^{15}\, mol\, Pa^{-1}} = 5.2_4 \times 10^{-11}\, Pa$$

With the fugacity in hand, we can now calculate the concentrations in each compartment:

$$\boxed{C_i = Z_i f_{Sys}}$$

$$C_A = Z_A f_{Sys} = 4.034_0 \times 10^{-4}\, mol\, Pa^{-1}\, m^{-3} \times 5.2_4 \times 10^{-11}\, Pa = 2.1_1 \times 10^{-14}\, mol\, m^{-3}$$

$$C_Q = Z_Q f_{Sys} = 1.1_4 \times 10^8\, mol\, Pa^{-1}\, m^{-3} \times 5.2_4 \times 10^{-11}\, Pa = 5.9_5 \times 10^{-3}\, mol\, m^{-3}$$

$$C_W = Z_W f_{Sys} = 21.6_0\, mol\, Pa^{-1}\, m^{-3} \times 5.2_4 \times 10^{-11}\, Pa = 1.1_3 \times 10^{-9}\, mol\, m^{-3}$$

$$C_{SS} = Z_{SS} f_{Sys} = 2.4_9 \times 10^6\, mol\, Pa^{-1}\, m^{-3} \times 5.2_4 \times 10^{-11}\, Pa = 1.3_0 \times 10^{-4}\, mol\, m^{-3}$$

$$C_{Fish} = Z_{Fish} f_{Sys} = 1.1_8 \times 10^6\, mol\, Pa^{-1}\, m^{-3} \times 5.2_4 \times 10^{-11}\, Pa = 6.2_0 \times 10^{-5}\, mol\, m^{-3}$$

$$C_{Soil} = Z_{Soil} f_{Sys} = 3.9_8 \times 10^5\, mol\, Pa^{-1}\, m^{-3} \times 5.2_4 \times 10^{-11}\, Pa = 2.0_8 \times 10^{-5}\, mol\, m^{-3}$$

$$C_{Sed} = Z_{Sed} f_{Sys} = 7.9_6 \times 10^5\, mol\, Pa^{-1}\, m^{-3} \times 5.2_4 \times 10^{-11}\, Pa = 4.1_7 \times 10^{-5}\, mol\, m^{-3}$$

The molar amounts in each compartment are given by CV:

$$\boxed{m_i = C_i V_i}$$

$$m_A = C_A V_A = 2.1_1 \times 10^{-14}\, mol\, m^{-3} \times 1.00_0 \times 10^{14}\, m^3 = 2.1_1\, mol$$

$$m_Q = C_Q V_Q = 5.9_5 \times 10^{-3}\, mol\, m^{-3} \times 2.00_0 \times 10^3\, m^3 = 11.9_0\, mol$$

$$m_W = C_W V_W = 1.1_3 \times 10^{-9}\, mol\, m^{-3} \times 2.00_0 \times 10^{11}\, m^3 = 2.2_6 \times 10^2\, mol$$

$$m_{SS} = C_{SS} V_{SS} = 1.3_0 \times 10^{-4}\, mol\, m^{-3} \times 1.00_0 \times 10^6\, m^3 = 1.3_0 \times 10^2\, mol$$

$$m_{Fish} = C_{Fish} V_{Fish} = 6.2_0 \times 10^{-5}\, mol\, m^{-3} \times 2.00_0 \times 10^5\, m^3 = 12._4\, mol$$

$$m_{Soil} = C_{Soil} V_{Soil} = 2.0_8 \times 10^{-5}\, mol\, m^{-3} \times 1.80 \times 10^{10}\, m^3 = 3.7_5 \times 10^5\, mol$$

$$m_{Sed} = C_{Sed} V_{Sed} = 4.1_7 \times 10^{-5}\, mol\, m^{-3} \times 5.00 \times 10^8\, m^3 = 2.0_8 \times 10^4\, mol$$

As a final check, we can sum the molar amounts and verify that they add up to the total chemical in the system, which is $3.963_3 \times 10^5\, mol$:

$$3.963_3 \times 10^5\, mol \stackrel{?}{=} \sum_{i=1}^{6} m_i$$

$$3.963_3 \times 10^5\, mol \stackrel{?}{=} \left(2.1_1 + 11.9_0 + 2.2_6 \times 10^2 + 1.3_0 \times 10^2 + 12._4 + 3.7_5 \times 10^5 + 2.0_8 \times 10^4\right) mol$$

$$3.963_3 \times 10^5\, mol = 3.9_6 \times 10^5\, mol$$

Level I RESULTS

Chemical Name Benzo[a]pyrene
Environment Name EQC- 2012-Hughes et al.

Fugacity Ratio	0.032841388
Subcooled Liq. Vap. Press. (Pa)	2.13146E-05
Fugacity (Pa)	5.238E-11

Partition Ratio	dimensionless
Air-Water	1.87E-05
Soil-Water	1.84E+04
Sediment-Water	3.68E+04
Suspended Particles-Water	1.15E+05
Fish-Water	5.48E+04
Aerosol-Water	-
Aerosol-Air	2.81E+11

	Z Value	**VZ**	**Concentration**			**Quantity**		
	mol/m³*Pa	mol/Pa	mol/m³	g/m³	ng/L	mol	kg	%
Air	4.0340E-04	4.0340E+10	2.113E-14	5.3313E-12	4.4427E-06	2.11	0.53	0.0
Aerosol	1.14E+08	2.2711E+11	5.95E-03	1.5007E+00	7.5037E+02	11.90	3.00	0.0
Water	2.160E+01	4.3196E+12	1.13E-09	2.8544E-07	2.8544E-04	226.26	57.09	0.1
Suspended Particles	2.49E+06	2.4866E+12	1.30E-04	3.2863E-02	2.1909E+01	130.25	32.86	0.0
Fish	1.18E+06	2.3682E+11	6.20E-05	1.5649E-02	1.5649E+01	12.40	3.13	0.0
Soil	3.98E+05	7.1614E+15	2.08E-05	5.2581E-03	2.1909E+00	375106.26	94645.31	94.6
Sediment	7.96E+05	3.9786E+14	4.17E-05	1.0516E-02	4.3817E+00	20839.24	5258.07	5.3
Total		7.57E+15				3.963E+05	100000.00	100

FIGURE 5.4 Screenshot of the Results screen from the Level I model, showing the calculation results for benzo[a]pyrene.

As a "bonus" for our efforts, we can also calculate the partition ratios between various compartments as Z_i/Z_j:

$$K_{AW} = \frac{Z_A}{Z_W} = \frac{4.034_0 \times 10^{-4} mol\,Pa^{-1}\,m^{-3}}{21.6_0 mol\,Pa^{-1}\,m^{-3}} = 1.86_8 \times 10^{-5}$$

$$K_{QA} = \frac{Z_Q}{Z_A} = \frac{1.1_4 \times 10^{8} mol\,Pa^{-1}\,m^{-3}}{4.034_0 \times 10^{-4} mol\,Pa^{-1}\,m^{-3}} = 2.8_1 \times 10^{11}$$

$$K_{Soil-W} = \frac{Z_{Soil}}{Z_W} = \frac{3.9_8 \times 10^{5} mol\,Pa^{-1}\,m^{-3}}{21.6_0 mol\,Pa^{-1}\,m^{-3}} = 1.8_4 \times 10^{4}$$

$$K_{Sed-W} = \frac{Z_{Sed}}{Z_W} = \frac{7.9_6 \times 10^{5} mol\,Pa^{-1}\,m^{-3}}{21.6_0 mol\,Pa^{-1}\,m^{-3}} = 3.6_8 \times 10^{4}$$

$$K_{SS-W} = \frac{Z_{SS}}{Z_W} = \frac{2.4_9 \times 10^{6} mol\,Pa^{-1}\,m^{-3}}{21.6_0 mol\,Pa^{-1}\,m^{-3}} = 1.1_5 \times 10^{5}$$

$$K_{Fish-W} = \frac{Z_{Fish}}{Z_W} = \frac{1.1_8 \times 10^{6} mol\,Pa^{-1}\,m^{-3}}{21.6_0 mol\,Pa^{-1}\,m^{-3}} = 5.4_8 \times 10^{4}$$

All these results may be verified by comparison with the Level I results output for this data, given in Figure 5.4.

5.2 LEVEL II MODELS

Level I models provide a useful, relatively simple means by which to "eyeball" an environmental modelling scenario, giving a rough picture of where a chemical will concentrate in the environment. A more sophisticated view is provided by the Level II approach, which integrates both advective transport and degradation losses into the analysis. The Level II approach assumes a steady-state condition between media that remain at equilibrium. As noted in Chapter 3, the main condition for determining the system fugacity in such a context is a mass balance based on rates of chemical entering and leaving the system, which leads to the following expression for the system fugacity:

$$f_{Sys} = \frac{I}{\sum_{i=1}^{n} D_i^L}$$

$$\sum_{i=1}^{n} D_i^L = \sum_{i=1}^{n} D_i^{Adv} + \sum_{i=1}^{n} D_i^{Deg}$$

The primary new task in constructing a Level II model is therefore to determine the various D-values for the loss processes, according to the expressions introduced in Chapter 3.

$$D_i^{Adv} = G_i^{Adv} Z_i$$

$$D_i^{Deg} = V_i Z_i k_i^{Deg}$$

As a working example, we will continue modelling the distribution of benzo[a]pyrene in the updated EQC environment (Hughes et al. 2012), using the Level II model from the CEMC website. Both the Level I and Level II versions use a seven-compartment approach and share the same assumptions for the fugacity capacities of each compartment, as outlined above for the Level I approach. Therefore, we can apply the same fugacity capacities that were derived for Level I in the Level II model calculations. Specifically:

$Z_A = 4.034_0 \times 10^{-4} \, mol\,Pa^{-1}\,m^{-3}$ $Z_Q = 1.1_4 \times 10^{8} \, mol\,Pa^{-1}\,m^{-3}$ $Z_W = 21.6_0 \, mol\,Pa^{-1}\,m^{-3}$

$Z_{SS} = 2.4_9 \times 10^{6} \, mol\,Pa^{-1}\,m^{-3}$ $Z_{Fish} = 1.1_8 \times 10^{6} \, mol\,Pa^{-1}\,m^{-3}$ $Z_{Soil} = 3.9_8 \times 10^{5} \, mol\,Pa^{-1}\,m^{-3}$

$Z_{Sed} = 7.9_6 \times 10^{5} \, mol\,Pa^{-1}\,m^{-3}$

The advective transport into air, water, and sediment is entered into the Level II calculation in terms of the advective flow residence time in each compartment. The default residence times from the Level II model are as shown in Figure 5.5.

As a result, before we can determine the D-values for advection, we need to calculate the advection rates for each compartment, where relevant. Recall the relationship between residence time and advection rate:

$$\tau_i = \frac{V_i Z_i}{D_i^{Adv}}$$

	Res. Time	
	hours	days
Air	100.0	4.167E+00
Aerosol	100.0	4.167E+00
Water	1000.0	4.167E+01
Suspended Particles	1000.0	4.167E+01
Fish	1000.0	4.167E+01
Soil	-	-
Sediment	50000.0	2.083E+03

FIGURE 5.5 Screenshot of the Advection Residence Times panel from the Chemical Properties screen from the Level II model available from the CEMC website, with input advective residence time values.

Solving for the needed D-value, we have:

$$\boxed{D_i^{Adv} = \frac{V_i Z_i}{\tau_i}}$$

The advection D-values for air, water, and sediment are therefore:

$$D_A^{Adv} = \frac{V_A Z_A}{\tau_A^{Adv}} = \frac{1.00_0 \times 10^{14} m^3 \times 4.034_0 \times 10^{-4} mol\, Pa^{-1}\, m^{-3}}{100.0\, h} = 4.03_4 \times 10^8 mol\, Pa^{-1}\, h^{-1}$$

$$D_Q^{Adv} = \frac{V_Q Z_Q}{\tau_Q^{Adv}} = \frac{2.00_0 \times 10^{3} m^3 \times 1.1_4 \times 10^{8} mol\, Pa^{-1}\, m^{-3}}{100.0\, h} = 2.2_7 \times 10^9 mol\, Pa^{-1}\, h^{-1}$$

$$D_W^{Adv} = \frac{V_W Z_W}{\tau_W^{Adv}} = \frac{2.00_0 \times 10^{11} m^3 \times 21.6_0 mol\, Pa^{-1}\, m^{-3}}{1000.0\, h} = 4.32_0 \times 10^9 mol\, Pa^{-1}\, h^{-1}$$

$$D_{SS}^{Adv} = \frac{V_{SS} Z_{SS}}{\tau_{SS}^{Adv}} = \frac{1.00_0 \times 10^{6} m^3 \times 2.4_9 \times 10^{6} mol\, Pa^{-1}\, m^{-3}}{1000.0\, h} = 2.4_9 \times 10^9 mol\, Pa^{-1}\, h^{-1}$$

$$D_{Fish}^{Adv} = \frac{V_{Fish} Z_{Fish}}{\tau_{Fish}^{Adv}} = \frac{2.00_0 \times 10^{5} m^3 \times 1.1_8 \times 10^{6} mol\, Pa^{-1}\, m^{-3}}{1000.0\, h} = 2.3_7 \times 10^8 mol\, Pa^{-1}\, h^{-1}$$

$$D_{Sed}^{Adv} = \frac{V_{Sed} Z_{Sed}}{\tau_{Sed}^{Adv}} = \frac{5.00 \times 10^{8} m^3 \times 7.9_6 \times 10^{5} mol\, Pa^{-1}\, m^{-3}}{50000.0\, h} = 7.9_6 \times 10^9 mol\, Pa^{-1}\, h^{-1}$$

The degradation rates are given in terms of reaction half-lives or half-times. The half-times and residence times are related through the rate constant as follows:

$$\frac{1}{\tau_i^{Deg}} = k_i^{Deg} = \frac{-\ln(0.5)}{\tau_i^{1/2-Deg}}$$

The relationship between degradation half-time and degradation D-value is therefore:

$$\boxed{D_i^{Deg} = \frac{-\ln(0.5) \times V_i Z_i}{\tau_i^{1/2-Deg}}}$$

The half-times for the various degradation processes are part of the chemical properties input and are shown in Figure 5.6 for this example.

	Half-Life	
	hours	days
Air	1.700E+02	7.083E+00
Aerosol	1.700E+02	7.083E+00
Water	1.700E+03	7.083E+01
Suspended Particles	1.700E+03	7.083E+01
Fish	2.677E+01	1.115E+00
Soil	1.700E+04	7.083E+02
Sediment	5.500E+04	2.292E+03

FIGURE 5.6 Screenshot of the Reaction Half-Lives panel from the Chemical Properties screen from the Level II model available from the CEMC website, with input degradation half-time values for benzo[a] pyrene.

Therefore, the D-values for degradation in the various compartments are:

$$D_A^{Deg} = \frac{-\ln(0.5) \times V_A Z_A}{\tau_A^{1/2-Deg}}$$

$$= \frac{-\ln(0.5) \times 1.00_0 \times 10^{14} m^3 \times 4.034_0 \times 10^{-4} mol\, Pa^{-1}\, m^{-3}}{170.00\, h} = 1.64_5 \times 10^8 mol\, Pa^{-1}\, h^{-1}$$

$$D_Q^{Deg} = \frac{-\ln(0.5) \times V_Q Z_Q}{\tau_Q^{1/2-Deg}}$$

$$= \frac{-\ln(0.5) \times 2.00_0 \times 10^3 m^3 \times 1.1_4 \times 10^8 mol\, Pa^{-1}\, m^{-3}}{170.00\, h} = 9.2_6 \times 10^8 mol\, Pa^{-1}\, h^{-1}$$

$$D_W^{Deg} = \frac{-\ln(0.5) \times V_W Z_W}{\tau_W^{1/2-Deg}}$$

$$= \frac{-\ln(0.5) \times 2.00_0 \times 10^{11} m^3 \times 21.6_0 mol\, Pa^{-1}\, m^{-3}}{1700.00\, h} = 1.76_1 \times 10^9 mol\, Pa^{-1}\, h^{-1}$$

$$D_{SS}^{Deg} = \frac{-\ln(0.5) \times V_{SS} Z_{SS}}{\tau_{SS}^{1/2-Deg}}$$

$$= \frac{-\ln(0.5) \times 1.00_0 \times 10^6 m^3 \times 2.4_9 \times 10^6 mol\, Pa^{-1}\, m^{-3}}{1700.00\, h} = 1.0_1 \times 10^9 mol\, Pa^{-1}\, h^{-1}$$

$$D_{Fish}^{Deg} = \frac{-\ln(0.5) \times V_{Fish} Z_{Fish}}{\tau_{Fish}^{1/2-Deg}}$$

$$= \frac{-\ln(0.5) \times 2.00_0 \times 10^5 m^3 \times 1.1_8 \times 10^6 mol\, Pa^{-1}\, m^{-3}}{26.77\, h} = 6.1_3 \times 10^9 mol\, Pa^{-1}\, h^{-1}$$

$$D_{Soil}^{Deg} = \frac{-\ln(0.5) \times V_{Soil} Z_{Soil}}{\tau_{Soil}^{1/2-Deg}}$$

$$= \frac{-\ln(0.5) \times 1.80 \times 10^{10} m^3 \times 3.9_8 \times 10^5 mol\, Pa^{-1}\, m^{-3}}{1.70 \times 10^4 h} = 2.9_2 \times 10^{11} mol\, Pa^{-1}\, h^{-1}$$

$$D_{Sed}^{Deg} = \frac{-\ln(0.5) \times V_{Sed} Z_{Sed}}{\tau_{Sed}^{1/2-Deg}}$$

$$= \frac{-\ln(0.5) \times 5.00 \times 10^8 m^3 \times 7.9_6 \times 10^5 mol\, Pa^{-1}\, m^{-3}}{55000.0\, h} = 5.0_1 \times 10^9 mol\, Pa^{-1}\, h^{-1}$$

The advantage of working with fugacity instead of concentration is now again clearly seen, in that the various D-values for these loss processes can be directly compared. A quick evaluation reveals the fact that degradation loss within the soil is by far the dominant loss process, being more than an order of magnitude greater than all other loss processes.

With these D-values in hand, we can now solve for the prevailing system fugacity:

$$\boxed{f_{Sys} = \frac{I}{\sum_{i=1}^{n} D_i^L}}$$

The sum of D-values for all losses from all compartments is:

$$D_i^L = D_i^{Adv} + D_i^{Deg}$$

$$D_i^L = \left(D_A^{Adv} + D_Q^{Adv} + D_W^{Adv} + D_{SS}^{Adv} + D_{Fish}^{Adv} + D_{Sed}^{Adv}\right) + \left(D_A^{Deg} + D_Q^{Deg} + D_W^{Deg} + D_{SS}^{Deg} + D_{Fish}^{Deg} + D_{Soil}^{Deg} + D_{Sed}^{Deg}\right)$$

$$D_i^L = \left(\begin{array}{c} \left(\begin{array}{c} 4.03_4 \times 10^8 + 2.2_7 \times 10^9 + 4.32_0 \times 10^9 \\ +2.4_9 \times 10^9 + 2.3_7 \times 10^8 + 7.9_6 \times 10^9 \end{array}\right) \\ + \left(\begin{array}{c} 1.64_5 \times 10^8 + 9.2_6 \times 10^8 + 1.76_1 \times 10^9 + 1.0_1 \times 10^9 \\ +6.1_3 \times 10^9 + 2.9_2 \times 10^{11} + 5.0_1 \times 10^9 \end{array}\right) \end{array}\right) mol\,Pa^{-1}\,h^{-1}$$

$$D_i^L = 3.2_5 \times 10^{11}\,mol\,Pa^{-1}\,h^{-1}$$

The default emission rate is given as $1.00 \times 10^5\ kg\ h^{-1}$ so we need to convert it to a molar emission rate as:

$$I = \frac{E\left(kg\,h^{-1}\right)}{M(g\,mol^{-1}) \times 10^{-3}\,kg\,g^{-1}} = \frac{1.00 \times 10^5\,kg\,h^{-1}}{252.316\,g\,mol^{-1} \times 10^{-3}\,kg\,g^{-1}} = 3.96_3 \times 10^5\,mol\,h^{-1}$$

The prevailing fugacity in the system is therefore:

$$f_{Sys} = \frac{I}{\sum_{i=1}^{n} D_i^L} = \frac{3.96_3 \times 10^5\,mol\,h^{-1}}{3.2_5 \times 10^{11}\,mol\,Pa^{-1}\,h^{-1}} = 1.2_2 \times 10^{-6}\,Pa$$

With the fugacity determined, we can now calculate the rates of the various loss processes to compare their magnitudes. The advective flow rates are given by:

$$\boxed{r_i^{Adv} = D_i^{Adv} f_{Sys}}$$

Therefore, for the six compartments with advection, we have the following rates:

$$r_A^{Adv} = D_A^{Adv} f_{Sys} = 4.034_0 \times 10^8\,mol\,Pa^{-1}\,h^{-1} \times 1.2_2 \times 10^{-6}\,Pa = 4.9_2 \times 10^2\,mol\,h^{-1}$$

$$r_Q^{Adv} = D_Q^{Adv} f_{Sys} = 2.2_7 \times 10^9\,mol\,Pa^{-1}\,h^{-1} \times 1.2_2 \times 10^{-6}\,Pa = 2.7_7 \times 10^3\,mol\,h^{-1}$$

$$r_W^{Adv} = D_W^{Adv} f_{Sys} = 4.32_0 \times 10^9\,mol\,Pa^{-1}\,h^{-1} \times 1.2_2 \times 10^{-6}\,Pa = 5.2_7 \times 10^3\,mol\,h^{-1}$$

$$r_{SS}^{Adv} = D_{SS}^{Adv} f_{Sys} = 2.4_9 \times 10^9\,mol\,Pa^{-1}\,h^{-1} \times 1.2_2 \times 10^{-6}\,Pa = 3.0_4 \times 10^3\,mol\,h^{-1}$$

$$r_{Fish}^{Adv} = D_{Fish}^{Adv} f_{Sys} = 2.3_7 \times 10^8\,mol\,Pa^{-1}\,h^{-1} \times 1.2_2 \times 10^{-6}\,Pa = 2.8_9 \times 10^2\,mol\,h^{-1}$$

$$r_{Sed}^{Adv} = D_{Sed}^{Adv} f_{Sys} = 7.9_6 \times 10^9\,mol\,Pa^{-1}\,h^{-1} \times 1.2_2 \times 10^{-6}\,Pa = 9.7_1 \times 10^3\,mol\,h^{-1}$$

The degradation loss rates are given by:

$$\boxed{r_i^{Deg} = D_i^{Deg} f_{Sys}}$$

Therefore, for the seven compartments, we have the following rates:

$$r_A^{Deg} = D_A^{Deg} f_{Sys} = 1.64_5 \times 10^8 \, mol \, Pa^{-1} \, h^{-1} \times 1.2_2 \times 10^{-6} \, Pa = 2.0_1 \times 10^2 \, mol \, h^{-1}$$

$$r_Q^{Deg} = D_Q^{Deg} f_{Sys} = 9.2_6 \times 10^8 \, mol \, Pa^{-1} \, h^{-1} \times 1.2_2 \times 10^{-6} \, Pa = 1.1_3 \times 10^3 \, mol \, h^{-1}$$

$$r_W^{Deg} = D_W^{Deg} f_{Sys} = 1.76_1 \times 10^9 \, mol \, Pa^{-1} \, h^{-1} \times 1.2_2 \times 10^{-6} \, Pa = 2.1_5 \times 10^3 \, mol \, h^{-1}$$

$$r_{SS}^{Deg} = D_{SS}^{Deg} f_{Sys} = 1.0_1 \times 10^9 \, mol \, Pa^{-1} \, h^{-1} \times 1.2_2 \times 10^{-6} \, Pa = 1.2_4 \times 10^3 \, mol \, h^{-1}$$

$$r_{Fish}^{Deg} = D_{Fish}^{Deg} f_{Sys} = 6.1_3 \times 10^9 \, mol \, Pa^{-1} \, h^{-1} \times 1.2_2 \times 10^{-6} \, Pa = 7.4_8 \times 10^3 \, mol \, h^{-1}$$

$$r_{Soil}^{Deg} = D_{Soil}^{Deg} f_{Sys} = 2.9_2 \times 10^{11} \, mol \, Pa^{-1} \, h^{-1} \times 1.2_2 \times 10^{-6} \, Pa = 3.5_6 \times 10^5 \, mol \, h^{-1}$$

$$r_{Sed}^{Deg} = D_{Sed}^{Deg} f_{Sys} = 5.0_1 \times 10^9 \, mol \, Pa^{-1} \, h^{-1} \times 1.2_2 \times 10^{-6} \, Pa = 6.1_2 \times 10^3 \, mol \, h^{-1}$$

It is clear that the dominant loss mechanism is chemical degradation in the soil compartment. However, these rates can be more easily compared by calculating the percentage of the total loss that is due to each process.

The total loss rate is the sum of all loss rates, which is:

$$r^L = \left(r_A^{Adv} + r_Q^{Adv} + r_W^{Adv} + r_{SS}^{Adv} + r_{Fish}^{Adv} + r_{Sed}^{Adv}\right) + \left(r_A^{Deg} + r_Q^{Deg} + r_W^{Deg} + r_{SS}^{Deg} + r_{Fish}^{Deg} + r_{Soil}^{Deg} + r_{Sed}^{Deg}\right)$$

$$r^L = 3.9_6 \times 10^5 \, mol \, h^{-1}$$

The percentage of the total loss rate due to each process is therefore:

$$\%r_A^{Adv} = \left(r_A^{Adv} / r_{Total}^L\right) \times 100\% = \left(4.9_2 \times 10^2 \, mol \, h^{-1} / 3.9_6 \times 10^5 \, mol \, h^{-1}\right) \times 100\% = 0.12\%$$

$$\%r_Q^{Adv} = \left(r_Q^{Adv} / r_{Total}^L\right) \times 100\% = \left(2.7_7 \times 10^3 \, mol \, h^{-1} / 3.9_6 \times 10^5 \, mol \, h^{-1}\right) \times 100\% = 0.70\%$$

$$\%r_W^{Adv} = \left(r_W^{Adv} / r_{Total}^L\right) \times 100\% = \left(5.2_7 \times 10^3 \, mol \, h^{-1} / 3.9_6 \times 10^5 \, mol \, h^{-1}\right) \times 100\% = 1.3\%$$

$$\%r_{SS}^{Adv} = \left(r_{SS}^{Adv} / r_{Total}^L\right) \times 100\% = \left(3.0_4 \times 10^3 \, mol \, h^{-1} / 3.9_6 \times 10^5 \, mol \, h^{-1}\right) \times 100\% = 0.77\%$$

$$\%r_{Fish}^{Adv} = \left(r_{Fish}^{Adv} / r_{Total}^L\right) \times 100\% = \left(2.8_9 \times 10^2 \, mol \, h^{-1} / 3.9_6 \times 10^5 \, mol \, h^{-1}\right) \times 100\% = 0.073\%$$

$$\%r_{Sed}^{Adv} = \left(r_{Sed}^{Adv} / r_{Total}^L\right) \times 100\% = \left(9.7_1 \times 10^3 \, mol \, h^{-1} / 3.9_6 \times 10^5 \, mol \, h^{-1}\right) \times 100\% = 2.5\%$$

The rates for losses due to degradation for all compartments are:

$$\%r_A^{Deg} = \left(r_A^{Deg} / r_{Total}^L\right) \times 100\% = \left(2.0_1 \times 10^2 \, mol \, h^{-1} / 3.9_6 \times 10^5 \, mol \, h^{-1}\right) \times 100\% = 0.051\%$$

$$\%r_Q^{Deg} = \left(r_Q^{Deg} / r_{Total}^L\right) \times 100\% = \left(1.1_3 \times 10^3 \, mol \, h^{-1} / 3.9_6 \times 10^5 \, mol \, h^{-1}\right) \times 100\% = 0.29\%$$

$$\%r_W^{Deg} = \left(r_W^{Deg} / r_{Total}^L\right) \times 100\% = \left(2.1_5 \times 10^3 \, mol \, h^{-1} / 3.9_6 \times 10^5 \, mol \, h^{-1}\right) \times 100\% = 0.54\%$$

$$\%r_{SS}^{Deg} = \left(r_{SS}^{Deg} / r_{Total}^L\right) \times 100\% = \left(1.2_4 \times 10^3 \, mol \, h^{-1} / 3.9_6 \times 10^5 \, mol \, h^{-1}\right) \times 100\% = 0.31\%$$

$$\%r_{Fish}^{Deg} = \left(r_{Fish}^{Deg} / r_{Total}^{L}\right) \times 100\% = \left(7.4_8 \times 10^3\, mol\, h^{-1} / 3.9_6 \times 10^5\, mol\, h^{-1}\right) \times 100\% = 1.9\%$$

$$\%r_{Soil}^{Deg} = \left(r_{Soil}^{Deg} / r_{Total}^{L}\right) \times 100\% = \left(3.5_6 \times 10^5\, mol\, h^{-1} / 3.9_6 \times 10^5\, mol\, h^{-1}\right) \times 100\% = 90.\%$$

$$\%r_{Sed}^{Deg} = \left(r_{Sed}^{Deg} / r_{Total}^{L}\right) \times 100\% = \left(6.1_2 \times 10^3\, mol\, h^{-1} / 3.9_6 \times 10^5\, mol\, h^{-1}\right) \times 100\% = 1.5\%$$

A quick check will show that these add up to 100% as they must, with about 95% loss due to reaction, and most of that is occurring in the soil where degradation is relatively fast. Clearly, the degradation in the soil is the main loss mechanism, and it will act as a loss "sink" for the system. The lifetime of the chemical in the system will be largely determined by this degradation rate, and it would be a regrettable error to invoke a significantly slower degradation rate from data obtained from another medium such as water. This would result in the conclusion that the chemical is much longer-lived or persistent in the environment than it is in reality, since the fastest degradation loss rate is the primary determinant of the overall loss rate for a system at equilibrium.

5.3 LEVEL III MODELS

The most sophisticated steady-state environmental model is an open system in which equilibrium is not assumed between any of the main compartments. In such a "Level III" model, each compartment generally has a unique fugacity, driving intercompartmental diffusive transport processes if physically possible. Equilibrium is assumed within a given compartment, such as for aerosols and pure air in the air medium; suspended solids, fish, and pure water in the water medium; pure air, pure water, and solid matter in the soil medium; and pure water and solid matter in sediment medium, are all assumed to be at equilibrium with the other components of their respective media.

We will continue to explore the EQC model environment and the corresponding Level III model computation for benzo[a]pyrene. Level III-type models involve intermedia transport, which is most conveniently done using bulk fugacity capacities to reduce the number of equations needed. Therefore, we will begin by calculating the bulk fugacity capacities of the air, water, soil, and sediment based on the *Z*-values already determined in the Level I section. To do so we require the volume fractions for each component, and the Level III default values are given in Figure 5.7.

The bulk fugacity capacity for the air medium is a volume-fraction-weighted sum of the fugacity capacities of air and aerosols:

$$Z_A^{Bulk} = v_A^{f-Q} Z_Q + v_A^{f-A} Z_A$$

$$Z_A^{Bulk} = 2.00 \times 10^{-11} \times 1.1_4 \times 10^8\, mol\, Pa^{-1}\, m^{-3} + \left(1 - 2.00 \times 10^{-11}\right) \times 4.034_0 \times 10^{-4}\, mol\, Pa^{-1}\, m^{-3}$$

$$Z_A^{Bulk} = 2.6_7 \times 10^{-3}\, mol\, Pa^{-1}\, m^{-3}$$

Volume Fractions

Aerosol in Air	2.00E-11
Susp. Particles in Water	5.00E-06
Fish in Water	1.00E-06
Air in Soil	0.200
Water in Soil	0.300
Solids in Soil	0.500
Water in Sediment	0.800
Solids in Sediment	0.200

FIGURE 5.7 Screenshot of the Volume Fractions panel from the Chemical Properties screen from the Level III model available from the CEMC website.

Note that, despite its vanishingly small volume fraction, the presence of aerosols in air has a significant impact on the fugacity capacity of bulk air. By volume, the fugacity capacity is about eleven orders of magnitude greater in aerosols than pure air, so the presence of aerosols greatly influences the capacity of air to hold a chemical.

The bulk fugacity capacity of water is obtained from a volume-fraction-weighted sum of the fugacity capacities of pure water, the suspended solids in water and fish.

$$Z_W^{Bulk} = v_W^{f-W} Z_W + v_W^{f-SS} Z_{SS} + v_W^{f-Fish} Z_{Fish}$$

$$Z_W^{Bulk} = \begin{bmatrix} \left(1-5.00\times10^{-6}-1.00\times10^{-6}\right)\times 21.6_0\, mol\, Pa^{-1}\, m^{-3} \\ +5.00\times10^{-6}\times 2.4_9\times10^{6}\, mol\, Pa^{-1}\, m^{-3} \\ +1.00\times10^{-6}\times 1.1_8\times10^{6}\, mol\, Pa^{-1}\, m^{-3} \end{bmatrix}$$

$$Z_W^{Bulk} = 35._2\, mol\, Pa^{-1}\, m^{-3}$$

The bulk soil fugacity is determined in a similar manner from the volume-fraction-weighted sum of the fugacity capacities of soil pore air, soil pore water, and soil solids.

$$Z_{Soil}^{Bulk} = v_{Soil}^{f-A} Z_A + v_{Soil}^{f-W} Z_W + v_{Soil}^{f-Solids} Z_{Soil}$$

$$Z_{Soil}^{Bulk} = \big[0.200\times 4.034_0\times10^{-4} mol\, Pa^{-1}\, m^{-3} + 0.300\times 21.6_0\, mol\, Pa^{-1}\, m^{-3} + 0.500\times 3.9_8\times10^{5}\, mol\, Pa^{-1}\, m^{-3}\big]$$

$$Z_{Soil}^{Bulk} = 1.9_9\times10^{5} mol\, Pa^{-1}\, m^{-3}$$

The bulk sediment fugacity follows from the volume-fraction-weighted sum of the fugacity capacities of sediment water and sediment solids:

$$Z_{Sed}^{Bulk} = v_{Sed}^{f-W} Z_W + v_{Sed}^{f-Solids} Z_{Sed}$$

$$Z_{Sed}^{Bulk} = \big[0.800\times 21.6_0\, mol\, Pa^{-1}\, m^{-3} + 0.200\times 7.9_6\times10^{5}\, mol\, Pa^{-1}\, m^{-3}\big]$$

$$Z_{Sed}^{Bulk} = 1.5_9\times10^{5} mol\, Pa^{-1}\, m^{-3}$$

With the bulk fugacity capacities in hand, we can now calculate the advective and degradation loss D-values for the bulk media as a whole:

$$D_A^{Adv} = \frac{V_A^{Bulk} Z_A^{Bulk}}{\tau_A^{Adv}} = \frac{1.00\times10^{14} m^3 \times 2.6_7\times10^{-3} mol\, Pa^{-1}\, m^{-3}}{100.0\, h} = 2.6_7\times10^{9} mol\, Pa^{-1}\, h^{-1}$$

$$D_W^{Adv} = \frac{V_W^{Bulk} Z_W^{Bulk}}{\tau_W^{Adv}} = \frac{2.00\times10^{11} m^3 \times 3.5_2\times10^{1}\, mol\, Pa^{-1}\, m^{-3}}{1000.0\, h} = 7.0_4\times10^{9} mol\, Pa^{-1}\, h^{-1}$$

$$D_{Sed}^{Adv} = \frac{V_{Sed}^{Bulk} Z_{Sed}^{Bulk}}{\tau_{Sed}^{Adv}} = \frac{5.00\times10^{8} m^3 \times 1.5_9\times10^{5} mol\, Pa^{-1}\, m^{-3}}{50000.0\, h} = 1.5_9\times10^{9} mol\, Pa^{-1}\, h^{-1}$$

The calculation of the degradation D-values is as before, if the half-times for each of the media components is the same, which is true for the air, soil, and sediment. For water, the calculation is somewhat more demanding, since the degradation half-times are not the same for the fish, water, or suspended solids components. In fact, the rate of degradation in the fish is much faster and will make a major contribution to the degradation D-value for water.

$$D_A^{Deg} = \frac{-\ln(0.5) \times V_A^{Bulk} Z_A^{Bulk}}{\tau_A^{1/2-Deg}}$$

$$= \frac{-\ln(0.5) \times 1.00 \times 10^{14} m^3 \times 2.6_7 \times 10^{-3} mol\,Pa^{-1}\,m^{-3}}{170.00\,h} = 1.0_9 \times 10^9 mol\,Pa^{-1}\,h^{-1}$$

$$D_{Soil}^{Deg} = \frac{-\ln(0.5) \times V_{Soil}^{Bulk} Z_{Soil}^{Bulk}}{\tau_{Soil}^{1/2-Deg}}$$

$$= \frac{-\ln(0.5) \times 1.80 \times 10^{10} m^3 \times 1.9_9 \times 10^5 mol\,Pa^{-1}\,m^{-3}}{1700.0\,h} = 1.4_6 \times 10^{11} mol\,Pa^{-1}\,h^{-1}$$

$$D_{Sed}^{Deg} = \frac{-\ln(0.5) \times V_{Sed}^{Bulk} Z_{Sed}^{Bulk}}{\tau_{Sed}^{1/2-Deg}}$$

$$= \frac{-\ln(0.5) \times 5.00 \times 10^8 m^3 \times 1.5_9 \times 10^5 mol\,Pa^{-1}\,m^{-3}}{55000.0\,h} = 1.0_0 \times 10^9 mol\,Pa^{-1}\,h^{-1}$$

For the water, we must treat degradation in the water, suspended solids, and fish as separate, parallel processes, for which we can ultimately sum the individual D-values:

$$D_W^{Deg-W} = \frac{-\ln(0.5) \times V_W^{Bulk} Z_W}{\tau_W^{1/2-Deg}}$$

$$= \frac{-\ln(0.5) \times 2.00 \times 10^{11} m^3 \times 21.6_0\, mol\,Pa^{-1}\,m^{-3}}{1700.00\,h} = 1.76_1 \times 10^9 mol\,Pa^{-1}\,h^{-1}$$

$$D_W^{Deg-SS} = \frac{-\ln(0.5) \times V_W^{Bulk} \times v_{SS}^f \times Z_{SS}}{\tau_{SS}^{1/2-Deg}}$$

$$= \frac{-\ln(0.5) \times 2.00 \times 10^{11} m^3 \times 5.00 \times 10^{-6} \times 2.4_9 \times 10^6\, mol\,Pa^{-1}\,m^{-3}}{1700.00\ h}$$

$$= 1.0_1 \times 10^9 mol\,Pa^{-1}\,h^{-1}$$

$$D_W^{Deg-Fish} = \frac{-\ln(0.5) \times V_W^{Bulk} \times v_{Fish}^f \times Z_{Fish}}{\tau_{Fish}^{1/2-Deg}}$$

$$= \frac{-\ln(0.5) \times 2.00 \times 10^{11} m^3 \times 1.00 \times 10^{-6} \times 1.1_8 \times 10^6\, mol\,Pa^{-1}\,m^{-3}}{26.77\,h}$$

$$= 6.1_3 \times 10^9 mol\,Pa^{-1}\,h^{-1}$$

Summing these three parallel contributions to the overall loss from degradation in the water, we arrive at:

$$D_W^{Deg} = D_W^{Deg-W} + D_{SS}^{Deg-W} + D_{Fish}^{Deg-W}$$

$$D_W^{Deg} = 1.76_1 \times 10^9 mol\,Pa^{-1}\,h^{-1} + 1.0_1 \times 10^9 mol\,Pa^{-1}\,h^{-1} + 6.1_3 \times 10^9 mol\,Pa^{-1}\,h^{-1}$$

$$D_W^{Deg} = 8.9_1 \times 10^9 mol\,Pa^{-1}\,h^{-1}$$

The next task is to determine the D-values for new intermedia diffusive transport, using the Whitman Two-Resistance approach. The new processes to be considered are bidirectional air–water, air–soil, and water–sediment diffusive transport. For these calculations, we require the necessary mass-transfer coefficients, the default values for which are given below.

Transport Processes	**Symbol**	**Mass-Transfer Coefficient ($m\ h^{-1}$)**
Air-side air–water	k_{AW}^{M-A}	5.00
Water-side air–water	k_{AW}^{M-W}	5.00×10^{-2}
Soil–Air boundary layer	$k_{Soil-A}^{Boundary}$	5.00
Soil pore air	k_{A-Soil}^{PoreA}	2.00×10^{-2}
Soil pore water	k_{Soil-W}^{PoreW}	1.00×10^{-5}
Dry deposition from air	U_Q^{Dry}	6.00×10^{-10}
Rain	U^{Rain}	1.00×10^{-4}
Scavenging ratio	$Q(unitless)$	2.00×10^{5}
Sediment deposition	U_{Sed-W}^{Dep}	5.00×10^{-7}
Sediment resuspension	U_{Sed-W}^{Resusp}	2.00×10^{-7}
Water runoff from soils	$U_{Soil-W}^{Runoff-W}$	5.00×10^{-5}
Solids runoff from soils	$U_{Soil-W}^{Runoff-Solids}$	1.00×10^{-8}

Interfacial areas used are as follows:

Interface	**Area (m^2)**
Air–Water	1.00×10^{10}
Air–Soil	9.00×10^{10}
Water–Sediment	1.00×10^{10}

For the air–water diffusive transport process, we have from Chapter 4:

$$\boxed{\frac{1}{D_{AW}^{Diff-Ov}} = \left(\frac{1}{D_{AW}^{Diff-A}} + \frac{1}{D_{AW}^{Diff-W}}\right)}$$

The two diffusive transport D-values follow from the corresponding mass-transfer coefficients, interfacial areas and fugacity capacities:

$$D_{AW}^{Diff-A} = k_{AW}^{M-A} A_{AW} Z_A$$

$$D_{AW}^{Diff-A} = 5.00\, m\, h^{-1} \times 1.00 \times 10^{10}\, m^2 \times 4.034_0 \times 10^{-4}\, mol\, Pa^{-1}\, m^{-3}$$

$$D_{AW}^{Diff-A} = 2.01_7 \times 10^{7}\, mol\, Pa^{-1}\, h^{-1}$$

$$D_{AW}^{Diff-W} = k_{AW}^{M-W} A_{AW} Z_W$$

$$D_{AW}^{Diff-W} = 5.00 \times 10^{-2}\, m\, h^{-1} \times 1.00 \times 10^{10}\, m^2 \times 21.6_0\, mol\, Pa^{-1}\, m^{-3}$$

$$D_{AW}^{Diff-W} = 1.08_0 \times 10^{10}\, mol\, Pa^{-1}\, h^{-1}$$

$$\frac{1}{D_{AW}^{Diff-Ov}} = \left(\frac{1}{D_{AW}^{Diff-A}} + \frac{1}{D_{AW}^{Diff-W}}\right)$$

$$\frac{1}{D_{AW}^{Diff-Ov}} = \left(\frac{1}{2.01_7 \times 10^7 \, mol\, Pa^{-1}\, h^{-1}} + \frac{1}{1.08_0 \times 10^{10} \, mol\, Pa^{-1}\, h^{-1}} \right)$$

$$\frac{1}{D_{AW}^{Diff-Ov}} = 4.96_7 \times 10^{-8} \, Pa\, h\, mol^{-1}$$

$$D_{AW}^{Diff-Ov} = 2.01_3 \times 10^7 \, mol\, Pa^{-1}\, h^{-1}$$

One can immediately see that the air-water diffusion is "air-side" controlled, as the overall D-value is essentially equal to the air-side D-value.

The soil-to-air diffusion D-value is determined in the manner introduced in Chapter 4 for soils, based on parallel diffusion through pore air and pore water, in series with diffusion through an air boundary layer above the soil surface. We then have:

$$\boxed{\frac{1}{D_{A-Soil}^{Diff-Ov}} = \frac{1}{D_{A-Soil}^{A-Boundary}} + \frac{1}{\left(D_{A-Soil}^{PoreA} + D_{Soil-W}^{PoreW}\right)}}$$

Where:

$$D_{A-Soil}^{A-Boundary} = k_{A-Soil}^{Boundary} A_{A-Soil} Z_A$$

$$D_{A-Soil}^{A-Boundary} = 5.00\, m\, h^{-1} \times 9.00 \times 10^{10}\, m^2 \times 4.034_0 \times 10^{-4}\, mol\, Pa^{-1}\, m^{-3}$$

$$D_{A-Soil}^{A-Boundary} = 1.81_5 \times 10^8 \, mol\, Pa^{-1}\, h^{-1}$$

$$D_{A-Soil}^{PoreA} = k_{A-Soil}^{PoreA} A_{Soil} Z_A$$

$$D_{A-Soil}^{PoreA} = 2.00 \times 10^{-2}\, m\, h^{-1} \times 9.00 \times 10^{10}\, m^2 \times 4.034_0 \times 10^{-4}\, mol\, Pa^{-1}\, m^{-3}$$

$$D_{A-Soil}^{PoreA} = 7.26_1 \times 10^5 \, mol\, Pa^{-1}\, h^{-1}$$

$$D_{Soil-W}^{PoreW} = k_{Soil-W}^{PoreW} A_{Soil} Z_W$$

$$D_{Soil-W}^{PoreW} = 1.00 \times 10^{-5}\, m\, h^{-1} \times 9.00 \times 10^{10}\, m^2 \times 21.6_0\, mol\, Pa^{-1}\, m^{-3}$$

$$D_{Soil-W}^{PoreW} = 1.94_4 \times 10^7 \, mol\, Pa^{-1}\, h^{-1}$$

Using these three D-values, we arrive at:

$$\frac{1}{D_{A-Soil}^{Diff-Ov}} = \frac{1}{D_{A-Soil}^{A-Boundary}} + \frac{1}{\left(D_{A-Soil}^{PoreA} + D_{Soil-W}^{PoreW}\right)}$$

$$\frac{1}{D_{A-Soil}^{Diff-Ov}} = \frac{1}{1.81_5 \times 10^8 \, mol\, Pa^{-1}\, h^{-1}} + \frac{1}{\left(7.26_1 \times 10^5 \, mol\, Pa^{-1}\, h^{-1} + 1.94_4 \times 10^7 \, mol\, Pa^{-1}\, h^{-1}\right)}$$

$$\frac{1}{D_{A-Soil}^{Diff-Ov}} = \frac{1}{1.81_5 \times 10^8 \, mol\, Pa^{-1}\, h^{-1}} + \frac{1}{2.01_7 \times 10^7 \, mol\, Pa^{-1}\, h^{-1}}$$

$$\frac{1}{D_{A-Soil}^{Diff-Ov}} = 5.51_0 \times 10^{-8} \, Pa\, h\, mol^{-1}$$

$$D_{A-Soil}^{Diff-Ov} = 1.81_5 \times 10^7 \, mol\, Pa^{-1}\, h^{-1}$$

Diffusive transport between sediment and water is treated in the Level III model simply as diffusion in water, with a D-value given by:

$$D_{Sed-W}^{Diff-Ov} = k_{Sed-W}^{Diff-W} A_{Sed-W} Z_W$$

$$D_{Sed-W}^{Diff-Ov} = 1.00 \times 10^{-4}\, m\, h^{-1} \times 1.00 \times 10^{10}\, m^2 \times 21.6_0\, mol\, Pa^{-1}\, m^{-3}$$

$$D_{Sed-W}^{Diff-Ov} = 2.16_0 \times 10^7\, mol\, Pa^{-1}\, h^{-1}$$

The Level III calculation includes a number of other transport processes for which we need to calculate D-values. These include aerosol and rain deposition on water, aerosol and rain deposition on soil, sediment deposition and resuspension, and runoff of soil water and soil solids to water. Such processes are often characterized by a measured U-value, an effective velocity in units of $m\ h^{-1}$ which, when multiplied by the area over which the process is occurring, gives a volume per hour rate. Note that sediment burial is accounted for as sediment advection, as in the Level II calculations.

For aerosol dry deposition on water, we have:

$$D_{AW}^{Dry} = U_Q^{Dry} A_{AW} v_A^{f-Q} Z_Q$$

$$D_{AW}^{Dry} = 6.00 \times 10^{-10}\, m\, h^{-1} \times 1.00 \times 10^{10}\, m^2 \times 2.00 \times 10^{-11} \times 1.1_4 \times 10^8\, mol\, Pa^{-1}\, m^{-3}$$

$$D_{AW}^{Dry} = 1.3_6 \times 10^{-2}\, mol\, Pa^{-1}\, h^{-1}$$

For aerosol wet deposition on water, we have:

$$D_{AW}^{Wet} = U^{Rain} Q A_{AW} v_A^{f-Q} Z_Q$$

$$D_{AW}^{Wet} = 1.00 \times 10^{-4}\, m\, h^{-1} \times 2.00 \times 10^5 \times 1.00 \times 10^{10}\, m^2 \times 2.00 \times 10^{-11} \times 1.1_4 \times 10^8\, mol\, Pa^{-1}\, m^{-3}$$

$$D_{AW}^{Wet} = 4.5_4 \times 10^8\, mol\, Pa^{-1}\, h^{-1}$$

The D-value for total aerosol deposition on water is simply the sum of the D-values for these two parallel processes:

$$D_{AW}^{Q} = D_{AW}^{Dry} + D_{AW}^{Wet}$$

$$D_{AW}^{Q} = 1.3_6 \times 10^{-2}\, mol\, Pa^{-1}\, h^{-1} + 4.5_4 \times 10^8\, mol\, Pa^{-1}\, h^{-1}$$

$$D_{AW}^{Q} = 4.5_4 \times 10^8\, mol\, Pa^{-1}\, h^{-1}$$

Clearly, dry deposition is a very minor process here, as its contribution does not change the total D-value from that of the wet deposition process.

For rain deposition on water, we have:

$$D_{AW}^{Rain} = U^{Rain} A_{AW} Z_W$$

$$D_{AW}^{Rain} = 1.00 \times 10^{-4}\, m\, h^{-1} \times 1.00 \times 10^{10}\, m^2 \times 21.6_0\, mol\, Pa^{-1}\, m^{-3}$$

$$D_{AW}^{Rain} = 2.16_0 \times 10^7\, mol\, Pa^{-1} h^{-1}$$

Aerosol deposition on soil is treated similarly to that on water.

For aerosol dry deposition on soil, we have:

$$D_{A-Soil}^{Dry} = U_Q^{Dry} A_{A-Soil} v_A^{f-Q} Z_Q$$

$$D_{A-Soil}^{Dry} = 6.00 \times 10^{-10}\, m\, h^{-1} \times 9.00 \times 10^{10}\, m^2 \times 2.00 \times 10^{-11} \times 1.1_4 \times 10^8\, mol\, Pa^{-1}\, m^{-3}$$

$$D_{A-Soil}^{Dry} = 1.2_3 \times 10^{-1}\, mol\, Pa^{-1}\, h^{-1}$$

For aerosol wet deposition on soil, we have:

$$D_{A-Soil}^{Wet} = U^{Rain} Q A_{A-Soil} v_A^{f-Q} Z_Q$$

$$D_{A-Soil}^{Wet} = 1.00 \times 10^{-4}\, m\, h^{-1} \times 2.00 \times 10^5 \times 9.00 \times 10^{10}\, m^2 \times 2.00 \times 10^{-11} \times 1.1_4 \times 10^8\, mol\, Pa^{-1}\, m^{-3}$$

$$D_{A-Soil}^{Wet} = 4.0_9 \times 10^9\, mol\, Pa^{-1}\, h^{-1}$$

The D-value for total aerosol deposition on soil is again the sum of the D-values for these two parallel processes:

$$D_{A-Soil}^{Q} = D_{A-Soil}^{Dry} + D_{A-Soil}^{Wet}$$

$$D_{A-Soil}^{Q} = 1.2_3 \times 10^{-1}\, mol\, Pa^{-1}\, h^{-1} + 4.0_9 \times 10^9\, mol\, Pa^{-1}\, h^{-1}$$

$$D_{A-Soil}^{Q} = 4.0_9 \times 10^9\, mol\, Pa^{-1}\, h^{-1}$$

Again, and as expected, dry deposition is a very minor process here.

For the rain deposition on soil, we have:

$$D_{A-Soil}^{Rain} = U^{Rain} A_{A-Soil} Z_W$$

$$D_{A-Soil}^{Rain} = 1.00 \times 10^{-4}\, m\, h^{-1} \times 9.00 \times 10^{10}\, m^2 \times 21.6_0\, mol\, Pa^{-1}\, m^{-3}$$

$$D_{A-Soil}^{Rain} = 1.94_4 \times 10^8\, mol\, Pa^{-1} h^{-1}$$

Sediment deposition and resuspension are similar:

$$D_{Sed-W}^{Dep} = U_{Sed-W}^{Dep} A_{Sed-W} Z_{SS}$$

$$D_{Sed-W}^{Dep} = 5.00 \times 10^{-7}\, m\, h^{-1} \times 1.00 \times 10^{10}\, m^2 \times 2.4_9 \times 10^6\, mol\, Pa^{-1}\, m^{-3}$$

$$D_{Sed-W}^{Dep} = 1.2_4 \times 10^{10}\, mol\, Pa^{-1} h^{-1}$$

$$D_{Sed-W}^{Resusp} = U_{Sed-W}^{Resusp} A_{Sed-W} Z_{Sed}$$

$$D_{Sed-W}^{Resusp} = 2.00 \times 10^{-7}\, m\, h^{-1} \times 1.00 \times 10^{10}\, m^2 \times 7.9_6 \times 10^5\, mol\, Pa^{-1}\, m^{-3}$$

$$D_{Sed-W}^{Resusp} = 1.5_9 \times 10^9\, mol\, Pa^{-1} h^{-1}$$

For soil–water runoff, we have:

$$D_{Soil-W}^{Runoff-W} = U_{Soil-W}^{Runoff-W} A_{A-Soil} Z_W$$

$$D_{Soil-W}^{Runoff-W} = 5.00 \times 10^{-5}\, m\, h^{-1} \times 9.00 \times 10^{10}\, m^2 \times 21.6_0\, mol\, Pa^{-1}\, m^{-3}$$

$$D_{Soil-W}^{Runoff-W} = 9.71_9 \times 10^7\, mol\, Pa^{-1} h^{-1}$$

Finally, for soil–solids runoff, we have:

$$D_{Soil-W}^{Runoff-Solids} = U_{Soil}^{Runoff-Solids} A_{A-Soil} Z_{Soil}$$

$$D_{Soil-W}^{Runoff-Solids} = 1.00 \times 10^{-8}\, m\, h^{-1} \times 9.00 \times 10^{10}\, m^2 \times 3.9_8 \times 10^5\, mol\, Pa^{-1}\, m^{-3}$$

$$D_{Soil-W}^{Runoff-Solids} = 3.5_8 \times 10^8\, mol\, Pa^{-1} h^{-1}$$

With these D-values calculated, we can now set out to solve for the various fugacities in each medium. From Chapter 4, we have the solutions to the fugacities already worked out (mercifully!):

$$f_W = \frac{\left[E_W + (J_1 J_4 / J_3) + \left(E_{Soil} D_{Soil-W}^{T-Soil} / D_{Soil}^{L}\right) + \left(E_{Sed} D_{Sed-W}^{T-Sed} / D_{Sed}^{L}\right)\right]}{\left[D_W^L - (J_2 J_4 / J_3) - \left(D_{Sed-W}^{T-W} D_{Sed-W}^{T-Sed} / D_{Sed}^{L}\right)\right]}$$

$$f_A = \frac{\left[J_1 + f_W J_2\right]}{J_3}$$

$$f_{Soil} = \frac{\left[E_{Soil} + f_A D_{A-Soil}^{T-A}\right]}{D_{Soil}^{L}}$$

$$f_{Sed} = \frac{\left[E_{Sed} + f_W D_{Sed-W}^{T-W}\right]}{D_{Sed}^{L}}$$

The various "Js" are given as:

$$J_1 = \frac{E_A}{D_A^L} + \frac{E_{Soil} D_{A-Soil}^{T-Soil}}{D_{Soil}^{L} D_A^L}$$

$$J_2 = \frac{D_{AW}^{T-W}}{D_A^L}$$

$$J_3 = 1 - \frac{D_{A-Soil}^{T-Soil} D_{A-Soil}^{T-A}}{D_A^L D_{Soil}^{L}}$$

$$J_4 = D_{AW}^{T} + \frac{D_{Soil-W}^{T-Soil} D_{A-Soil}^{T-A}}{D_{Soil}^{L}}$$

To use these, we need to know the total D-values for transport between various pairs of media, as well as the sum of losses for each medium.

Beginning with the total for transport from air to water:

$$D_{AW}^{T-A} = D_{AW}^{Diff-Ov} + D_{AW}^{Rain} + D_{AW}^{Q}$$

$$D_{AW}^{T-A} = \left(2.01_3 \times 10^7 + 2.16_0 \times 10^7 + 4.5_4 \times 10^8\right) mol\, Pa^{-1}\, h^{-1}$$

$$D_{AW}^{T-A} = 4.9_6 \times 10^8\, mol\, Pa^{-1}\, h^{-1}$$

Total transport D-value from air to soil:

$$D_{A-Soil}^{T-A} = D_{A-Soil}^{Diff-Ov} + D_{A-Soil}^{Rain} + D_{A-Soil}^{Q}$$

$$D_{A-Soil}^{T-A} = \left(1.81_5 \times 10^7 + 1.94_4 \times 10^8 + 4.0_9 \times 10^9\right) mol\, Pa^{-1}\, h^{-1}$$

$$D_{A-Soil}^{T-A} = 4.3_0 \times 10^9\, mol\, Pa^{-1}\, h^{-1}$$

The total transport D-values for soil to air is simply due to diffusive transport:

$$D_{A-Soil}^{T-Soil} = D_{A-Soil}^{Diff-Ov}$$

$$D_{A-Soil}^{T-Soil} = 1.81_5 \times 10^7 \, mol\, Pa^{-1}\, h^{-1}$$

The total for transport from water to air is determined solely by diffusive transport:

$$D_{AW}^{T-W} = D_{AW}^{Diff-Ov} = 2.01_3 \times 10^7 \, mol\, Pa^{-1}\, h^{-1}$$

Soil–water transport is solely by water and solids runoff:

$$D_{Soil-W}^{T-Soil} = D_{Soil-W}^{Runoff-W} + D_{Soil-W}^{Runoff-Solids}$$

$$D_{Soil-W}^{T-Soil} = 9.71_9 \times 10^7 \, mol\, Pa^{-1} h^{-1} + 3.5_8 \times 10^8 \, mol\, Pa^{-1} h^{-1}$$

$$D_{Soil-W}^{T-Soil} = 4.5_5 \times 10^8 \, mol\, Pa^{-1} h^{-1}$$

The total transport D-values for water to sediment is given by:

$$D_{Sed-W}^{T-W} = D_{Sed-W}^{Diff-Ov} + D_{Sed-W}^{Dep}$$

$$D_{Sed-W}^{T-W} = 2.16_0 \times 10^7 \, mol\, Pa^{-1}\, h^{-1} + 1.2_4 \times 10^{10} \, mol\, Pa^{-1} h^{-1}$$

$$D_{Sed-W}^{T-W} = 1.2_5 \times 10^{10} \, mol\, Pa^{-1} h^{-1}$$

The total transport D-values for sediment to water is given by:

$$D_{Sed-W}^{T-Sed} = D_{Sed-W}^{Diff-Ov} + D_{Sed-W}^{Resusp}$$

$$D_{Sed-W}^{T-Sed} = 2.16_0 \times 10^7 \, mol\, Pa^{-1}\, h^{-1} + 1.5_9 \times 10^9 \, mol\, Pa^{-1} h^{-1}$$

$$D_{Sed-W}^{T-Sed} = 1.6_1 \times 10^9 \, mol\, Pa^{-1} h^{-1}$$

The sums of losses for each medium are:

$$D_A^L = D_A^{Adv} + D_A^{Deg} + D_{AW}^{Diff-Ov} + D_{AW}^{Rain} + D_{AW}^{Q} + D_{A-Soil}^{Diff-Ov} + D_{A-Soil}^{Rain} + D_{A-Soil}^{Q}$$

$$D_A^L = \begin{pmatrix} 2.6_7 \times 10^9 + 1.0_9 \times 10^9 + 2.01_3 \times 10^7 + 2.16_0 \times 10^7 \\ +4.5_4 \times 10^8 + 1.81_5 \times 10^7 + 1.94_4 \times 10^8 + 4.0_9 \times 10^9 \end{pmatrix} mol\, Pa^{-1}\, h^{-1}$$

$$D_A^L = 8.5_6 \times 10^9 \, mol\, Pa^{-1} h^{-1}$$

$$D_W^L = D_W^{Adv} + D_W^{Deg} + D_{AW}^{Diff-Ov} + D_{Sed-W}^{Diff-Ov} + D_{Sed-W}^{Dep}$$

$$D_W^L = \left(7.0_4 \times 10^9 + 8.9_1 \times 10^9 + 2.01_3 \times 10^7 + 2.16_0 \times 10^7 + 1.2_4 \times 10^{10}\right) mol\, Pa^{-1}\, h^{-1}$$

$$D_W^L = 2.8_4 \times 10^{10} \, mol\, Pa^{-1} h^{-1}$$

$$D_{Soil}^L = D_{Soil}^{Deg} + D_{Soil-A}^{Diff-Ov} + D_{Soil-W}^{Runoff-W} + D_{Soil-W}^{Runoff-Solids}$$

$$D_{Soil}^L = \left(1.4_6 \times 10^{11} + 1.81_5 \times 10^7 + 9.71_9 \times 10^7 + 3.5_8 \times 10^8\right) mol\, Pa^{-1} h^{-1}$$

$$D_{Soil}^L = 1.4_6 \times 10^{11} \, mol\, Pa^{-1} h^{-1}$$

$$D^L_{Sed} = D^{Adv}_{Sed} + D^{Deg}_{Sed} + D^{Diff-Ov}_{Sed-W} + D^{Resusp}_{Sed-W}$$

$$D^L_{Sed} = \left(1.5_9 \times 10^9 + 1.0_0 \times 10^9 + 2.16_0 \times 10^7 + 1.5_9 \times 10^9\right) mol\, Pa^{-1} h^{-1}$$

$$D^L_{Sed} = 4.2_1 \times 10^9\, mol\, Pa^{-1} h^{-1}$$

The default emissions in the Level III model are 1.00×10^3 $kg\ h^{-1}$ in each of the air, water, and soil media, with no emission to the sediment. This mass-based rate must be converted to a molar flow rate, according to the following equation:

$$E_i(mol\ h^{-1}) = \frac{E_i\left(kg\, h^{-1}\right)}{M\,(g\, mol^{-1}) \times 10^{-3}\, kg\, g^{-1}} = \frac{1.00 \times 10^3\, kg\, h^{-1}}{252.316\, g\, mol^{-1} \times 10^{-3}\, kg\, g^{-1}} = 3.96_3 \times 10^3\, mol\, h^{-1}$$

Finally, we are ready to calculate the fugacities. We begin by evaluating the "J" values from the Mackay et al. (1992) solution to calculate the Level III mass-balance equations.

$$J_1 = \frac{E_A}{D^L_A} + \frac{E_{Soil} D^{T-Soil}_{A-Soil}}{D^L_{Soil} D^L_A}$$

$$J_1 = \frac{3.96_3 \times 10^3\, mol\, h^{-1}}{8.5_6 \times 10^9\, mol\, Pa^{-1}\, h^{-1}} + \frac{3.96_3 \times 10^3\, mol\, h^{-1} \times 1.81_5 \times 10^7\, mol\, Pa^{-1}\, h^{-1}}{1.4_6 \times 10^{11}\, mol\, Pa^{-1}\, h^{-1} \times 8.5_6 \times 10^9\, mol\, Pa^{-1}\, h^{-1}}$$

$$J_1 = 4.6_3 \times 10^{-7}\, Pa$$

$$J_2 = \frac{D^{T-W}_{AW}}{D^L_A}$$

$$J_2 = \frac{2.01_3 \times 10^7\, mol\, Pa^{-1} h^{-1}}{8.5_6 \times 10^9\, mol\, Pa^{-1}\, h^{-1}}$$

$$J_2 = 2.3_5 \times 10^{-3}$$

$$J_3 = 1 - \frac{D^{T-Soil}_{A-Soil} D^{T-A}_{A-Soil}}{D^L_A D^L_{Soil}}$$

$$J_3 = 1 - \left(\frac{1.81_5 \times 10^7\, mol\, Pa^{-1}\, h^{-1} \times 4.3_0 \times 10^9\, mol\, Pa^{-1}\, h^{-1}}{8.5_6 \times 10^9\, mol\, Pa^{-1}\, h^{-1} \times 1.4_6 \times 10^{11}\, mol\, Pa^{-1}\, h^{-1}}\right)$$

$$J_3 = 1.00$$

$$J_4 = D^{T-A}_{AW} + \frac{D^{T-Soil}_{A-Soil} D^{T-A}_{A-Soil}}{D^L_{Soil}}$$

$$J_4 = \left(4.9_6 \times 10^8 + \left[\frac{4.5_5 \times 10^8 \times 4.3_0 \times 10^9}{1.4_6 \times 10^{11}}\right]\right) mol\, Pa^{-1} h^{-1}$$

$$J_4 = 5.0_9 \times 10^8\, mol\, Pa^{-1} h^{-1}$$

Now evaluating the media fugacities:

$$f_W = \frac{\begin{bmatrix} E_W + (J_1 J_4 / J_3) \\ +\left(E_{Soil} D_{Soil-W}^{T-Soil} / D_{Soil}^{L}\right) \\ +\left(E_{Sed} D_{Sed-W}^{T-Sed} / D_{Sed}^{L}\right) \end{bmatrix}}{\begin{bmatrix} D_W^L - (J_2 J_4 / J_3) \\ -\left(D_{Sed-W}^{T-W} D_{Sed-W}^{T-Sed} / D_{Sed}^{L}\right) \end{bmatrix}}$$

$$f_W = \frac{\begin{bmatrix} 3.96_3 \times 10^3\, mol\, h^{-1} + \left(4.6_3 \times 10^{-7}\, Pa \times 5.0_9 \times 10^8\, mol\, Pa^{-1} h^{-1} / 1.00\right) \\ +\left(3.96_3 \times 10^3\, mol\, h^{-1} \times 4.5_5 \times 10^8\, mol\, Pa^{-1} h^{-1} / 1.4_6 \times 10^{11}\, mol\, Pa^{-1} h^{-1}\right) \\ +\left(0\, mol\, h^{-1} \times 1.6_1 \times 10^9\, mol\, Pa^{-1} h^{-1} / 4.2_1 \times 10^9\, mol\, Pa^{-1} h^{-1}\right) \end{bmatrix}}{\begin{bmatrix} 2.8_4 \times 10^{10}\, mol\, Pa^{-1} h^{-1} - \left(2.3_5 \times 10^{-3} \times 5.0_9 \times 10^8\, mol\, Pa^{-1} h^{-1} / 1.00\right) \\ -\left(1.2_5 \times 10^{10}\, mol\, Pa^{-1} h^{-1} \times 1.6_1 \times 10^9\, mol\, Pa^{-1} h^{-1} / 4.2_1 \times 10^9\, mol\, Pa^{-1} h^{-1}\right) \end{bmatrix}}$$

$$f_W = \frac{4.2_4 \times 10^3\, mol\, h^{-1}}{2.3_6 \times 10^{10}\, mol\, Pa^{-1}\, h^{-1}}$$

$$f_W = 1.7_8 \times 10^{-7}\, Pa$$

$$f_A = \frac{\left[J_1 + f_W J_2\right]}{J_3}$$

$$f_A = \frac{\left[4.6_3 \times 10^{-7}\, Pa + 1.7_8 \times 10^{-7}\, Pa \times 2.3_5 \times 10^{-3}\right]}{1.00}$$

$$f_A = 4.6_3 \times 10^{-7}\, Pa$$

$$f_{Soil} = \frac{\left[E_{Soil} + f_A D_{A-Soil}^{T-A}\right]}{D_{Soil}^{L}}$$

$$f_{Soil} = \frac{\left[3.96_3 \times 10^3\, mol\, h^{-1} + 4.6_3 \times 10^{-7}\, Pa \times 4.3_0 \times 10^9\, mol\, Pa^{-1} h^{-1}\right]}{1.4_6 \times 10^{11}\, mol\, Pa^{-1} h^{-1}}$$

$$f_{Soil} = 4.0_7 \times 10^{-8}\, Pa$$

$$f_{Sed} = \frac{\left[E_{Sed} + f_W D_{Sed-W}^{T-W}\right]}{D_{Sed}^{L}}$$

$$f_{Sed} = \frac{\left[0.00\, mol\, h^{-1} + 1.7_8 \times 10^{-7}\, Pa \times 1.2_5 \times 10^{10}\, mol\, Pa^{-1} h^{-1}\right]}{4.2_1 \times 10^9\, mol\, Pa^{-1} h^{-1}}$$

$$f_{Sed} = 5.2_7 \times 10^{-7}\, Pa$$

Having finally arrived at the fugacities for the four media, we can now do some calculations to compare concentrations and amounts, as well as rates of different processes. Ultimately, the aim is to reveal where the chemical concentrates and which transport processes are dominant in getting it there.

Concentrations follow from Zf products, each medium and its components having a unique fugacity.

The concentrations in the bulk air and its components (specified here as superscripts) are:

$$C_A^{Bulk} = Z_A^{Bulk} f_A = 2.6_7 \times 10^{-3} mol\, Pa^{-1} m^{-3} \times 4.6_3 \times 10^{-7} Pa = 1.2_4 \times 10^{-9} mol\, m^{-3}$$

$$C_A^A = Z_A^A f_A = 4.034_0 \times 10^{-4} mol\, Pa^{-1} m^{-3} \times 4.6_3 \times 10^{-7} Pa = 1.8_7 \times 10^{-10} mol\, m^{-3}$$

$$C_A^Q = Z_A^Q f_A = 1.1_4 \times 10^{8} mol\, Pa^{-1} m^{-3} \times 4.6_3 \times 10^{-7} Pa = 52._6 mol\, m^{-3}$$

Here, the concentration in the aerosols is more than eleven orders of magnitude greater compared to the gaseous air itself. Despite the very low volume fraction of the aerosols, their contribution has increased the concentration in the bulk air by nearly an order of magnitude with respect to gaseous air.

The bulk water and its components are:

$$C_W^{Bulk} = Z_W^{Bulk} f_W = 35._2 mol\, Pa^{-1} m^{-3} \times 1.7_8 \times 10^{-7} Pa = 6.2_7 \times 10^{-6} mol\, m^{-3}$$

$$C_W^W = Z_W^W f_W = 21.6_0 mol\, Pa^{-1} m^{-3} \times 1.7_8 \times 10^{-7} Pa = 3.8_5 \times 10^{-6} mol\, m^{-3}$$

$$C_W^{SS} = Z_W^{SS} f_W = 2.4_9 \times 10^{6} mol\, Pa^{-1} m^{-3} \times 1.7_8 \times 10^{-7} Pa = 0.44_3 mol\, m^{-3}$$

$$C_W^{Fish} = Z_W^{Fish} f_W = 1.1_8 \times 10^{6} mol\, Pa^{-1} m^{-3} \times 1.7_8 \times 10^{-7} Pa = 0.21_1 mol\, m^{-3}$$

Note that the much greater fugacity capacities of the suspended solids and the fish result in about five orders of magnitude increase in concentration in these water components compared to pure water.

The concentrations in the bulk soil and its components follow. Recall that the fugacity capacity of soil in Level I and II was that of the organic matter, which is separate from the pore air and water in soil within Level III:

$$C_{Soil}^{Bulk} = Z_{Soil}^{Bulk} f_{Soil} = 1.9_9 \times 10^{5} mol\, Pa^{-1} m^{-3} \times 4.0_7 \times 10^{-8} Pa = 8.0_9 \times 10^{-3} mol\, m^{-3}$$

$$C_{Soil}^A = Z_{Soil}^A f_{Soil} = 4.034_0 \times 10^{-4} mol\, Pa^{-1} m^{-3} \times 4.0_7 \times 10^{-8} Pa = 1.6_4 \times 10^{-11} mol\, m^{-3}$$

$$C_{Soil}^W = Z_{Soil}^W f_{Soil} = 21.6_0 mol\, Pa^{-1} m^{-3} \times 4.0_7 \times 10^{-8} Pa = 8.7_8 \times 10^{-7} mol\, m^{-3}$$

$$C_{Soil}^{OM} = Z_{Soil}^{OM} f_{Soil} = 3.9_8 \times 10^{5} mol\, Pa^{-1} m^{-3} \times 4.0_7 \times 10^{-8} Pa = 1.6_2 \times 10^{-2} mol\, m^{-3}$$

Here again, the concentration is much greater in the organic matter in the soil, reflecting, in part, the high K_{OW} value for benzo[a]pyrene (log K_{OW} = 6.04). The same effect is seen in the sediment concentrations below.

The bulk sediment concentration, and that of its components, are:

$$C_{Sed}^{Bulk} = Z_{Sed}^{Bulk} f_{Sed} = 1.5_9 \times 10^{5} mol\, Pa^{-1} m^{-3} \times 5.2_7 \times 10^{-7} Pa = 8.3_9 \times 10^{-2} mol\, m^{-3}$$

$$C_{Sed}^W = Z_{Sed}^W f_{Sed} = 21.6_0 mol\, Pa^{-1} m^{-3} \times 5.2_7 \times 10^{-7} Pa = 1.1_4 \times 10^{-5} mol\, m^{-3}$$

$$C_{Sed}^{OM} = Z_{Sed}^{OM} f_{Sed} = 7.9_6 \times 10^{5} mol\, Pa^{-1} m^{-3} \times 5.2_7 \times 10^{-7} Pa = 0.41_9 mol\, m^{-3}$$

What emerges is the significant concentration of benzo[a]pyrene in organic matter and biota, typical of a chemical with a high log K_{OW} and a low log K_{AW} (log $(Z_A/Z_W) = -4.73$). Benzo[a]pyrene is very hydrophobic, but it is also "aerophobic", resulting in its concentration in any medium that exhibits solvation properties resembling octanol, i.e., lipid- or organic carbon-rich media.

Concentrations of chemicals in a given compartment do not necessarily indicate where most of the chemical is to be found, since the amount of a chemical in a compartment is given not by the concentration alone, but the product of the concentration and the compartment volume. Thus, a high concentration of a chemical in a compartment with a vanishingly small volume is likely of less consequence than a moderate concentration of a chemical in a large-volume compartment. An obvious exception to this general statement concerns biota that are part of a food chain, which may have a small relative volume yet contain high concentrations of chemical.

Total amounts of benzo[a]pyrene in the EQC environment are given by:

$$m_A = C_A^{Bulk} V_A = 1.2_4 \times 10^{-9} mol\, m^{-3} \times 1.00_0 \times 10^{14}\, m^3 = 1.2_4 \times 10^5 mol$$

$$m_W = C_W^{Bulk} V_W = 6.2_7 \times 10^{-6} mol\, m^{-3} \times 2.00_0 \times 10^{11}\, m^3 = 1.2_5 \times 10^6 mol$$

$$m_{Soil} = C_{Soil}^{Bulk} V_{Soil} = 8.0_9 \times 10^{-3} mol\, m^{-3} \times 1.80 \times 10^{10}\, m^3 = 1.4_6 \times 10^8 mol$$

$$m_{Sed} = C_{Sed}^{Bulk} V_{Sed} = 8.3_9 \times 10^{-2} mol\, m^{-3} \times 5.00 \times 10^8\, m^3 = 4.2_0 \times 10^7 mol$$

From these results, it is clear that the great majority of the benzo[a]pyrene is in the soil and sediment. The fact that no emission is occurring directly to the sediment demonstrates the importance of water-to-sediment transfer, mostly by way of net sediment deposition from water.

We can also investigate the relative rates of the various transport processes involved in the Level III model using the EQC environment. Rates of processes follow directly from $r = Df$.

The advection rates are:

$$r_A^{Adv} = D_A^{Adv} f_A = 2.6_7 \times 10^9 mol\, Pa^{-1}\, h^{-1} \times 4.6_3 \times 10^{-7} Pa = 1.2_4 \times 10^3 mol\, h^{-1}$$

$$r_W^{Adv} = D_W^{Adv} f_W = 7.0_4 \times 10^9 mol\, Pa^{-1}\, h^{-1} \times 1.7_8 \times 10^{-7} Pa = 1.2_5 \times 10^3 mol\, h^{-1}$$

$$r_{Sed}^{Adv} = D_{Sed}^{Adv} f_{Sed} = 1.5_9 \times 10^9 mol\, Pa^{-1}\, h^{-1} \times 5.2_7 \times 10^{-7} Pa = 8.3_9 \times 10^2 mol\, h^{-1}$$

The degradation rates are:

$$r_A^{Deg} = D_A^{Deg} f_A = 1.0_9 \times 10^9 mol\, Pa^{-1}\, h^{-1} \times 4.6_3 \times 10^{-7} Pa = 5.0_5 \times 10^2 mol\, h^{-1}$$

$$r_W^{Deg} = D_W^{Deg} f_W = 8.9_1 \times 10^9 mol\, Pa^{-1}\, h^{-1} \times 1.7_8 \times 10^{-7} Pa = 1.5_9 \times 10^3 mol\, h^{-1}$$

$$r_{Soil}^{Deg} = D_{Soil}^{Deg} f_{Soil} = 1.4_6 \times 10^{11} mol\, Pa^{-1}\, h^{-1} \times 4.0_7 \times 10^{-8} Pa = 5.9_4 \times 10^3 mol\, h^{-1}$$

$$r_{Sed}^{Deg} = D_{Sed}^{Deg} f_{Sed} = 1.0_0 \times 10^9 mol\, Pa^{-1}\, h^{-1} \times 5.2_7 \times 10^{-7} Pa = 5.2_9 \times 10^2 mol\, h^{-1}$$

Intermedia diffusive transfer rates are given by:

$$r_{AW}^{Diff-A} = D_{AW}^{Diff-Ov} f_A = 2.01_3 \times 10^7\, mol\, Pa^{-1}\, m^{-3} \times 4.6_3 \times 10^{-7} Pa = 9.33_0\, mol\, h^{-1}$$

$$r_{AW}^{Diff-W} = D_{AW}^{Diff-Ov} f_W = 2.01_3 \times 10^7 mol\, Pa^{-1}\, h^{-1} \times 1.7_8 \times 10^{-7} Pa = 3.58_5\, mol\, h^{-1}$$

$$r_{A-Soil}^{Diff-Soil} = D_{A-Soil}^{Diff-Ov} f_{Soil} = 1.81_5 \times 10^7 mol\, Pa^{-1}\, h^{-1} \times 4.0_7 \times 10^{-8} Pa = 0.73_8 mol\, h^{-1}$$

$$r_{A-Soil}^{Diff-A} = D_{A-Soil}^{Diff-Ov} f_A = 1.81_5 \times 10^7 mol\, Pa^{-1}\, h^{-1} \times 4.6_3 \times 10^{-7} Pa = 0.84_1\, mol\, h^{-1}$$

$$r_{Sed-W}^{Diff-Sed} = D_{Sed-W}^{Diff-Ov} f_{Sed} = 2.16_0 \times 10^7 \, mol \, Pa^{-1} \, h^{-1} \times 5.2_7 \times 10^{-7} \, Pa = 11._4 \, mol \, h^{-1}$$

$$r_{Sed-W}^{Diff-W} = D_{Sed-W}^{Diff-Ov} f_W = 2.16_0 \times 10^7 \, mol \, Pa^{-1} \, h^{-1} \times 1.7_8 \times 10^{-7} \, Pa = 3.8_5 \, mol \, h^{-1}$$

Rain and aerosol deposition rates are:

$$r_{AW}^{Rain} = D_{AW}^{Rain} f_A = 2.16_0 \times 10^7 \, mol \, Pa^{-1} h^{-1} \times 4.6_3 \times 10^{-7} \, Pa = 10._0 \, mol \, h^{-1}$$

$$r_{A-Soil}^{Rain} = D_{A-Soil}^{Rain} f_A = 1.94_4 \times 10^8 \, mol \, Pa^{-1} h^{-1} \times 4.6_3 \times 10^{-7} \, Pa = 90._1 \, mol \, h^{-1}$$

$$r_{AW}^{Q} = D_{AW}^{Q} f_A = 4.5_4 \times 10^8 \, mol \, Pa^{-1} \, h^{-1} \times 4.6_3 \times 10^{-7} \, Pa = 2.1_0 \times 10^2 \, mol \, h^{-1}$$

$$r_{A-Soil}^{Q} = D_{A-Soil}^{Q} f_A = 4.0_9 \times 10^9 \, mol \, Pa^{-1} \, h^{-1} \times 4.6_3 \times 10^{-7} \, Pa = 1.8_9 \times 10^3 \, mol \, h^{-1}$$

Sediment deposition and resuspension rates are:

$$r_{Sed-W}^{Resusp} = D_{Sed-W}^{Resusp} f_{Sed} = 1.5_9 \times 10^9 \, mol \, Pa^{-1} \, h^{-1} \times 5.2_7 \times 10^{-7} \, Pa = 8.3_9 \times 10^2 \, mol \, h^{-1}$$

$$r_{Sed-W}^{Dep} = D_{Sed-W}^{Dep} f_W = 1.2_4 \times 10^{10} \, mol \, Pa^{-1} \, h^{-1} \times 1.7_8 \times 10^{-7} \, Pa = 2.2_1 \times 10^3 \, mol \, h^{-1}$$

Soil runoff rates are:

$$r_{Soil-W}^{Runoff-W} = D_{Soil-W}^{Runoff-W} f_{Soil} = 9.71_9 \times 10^7 \, mol \, Pa^{-1} h^{-1} \times 4.0_7 \times 10^{-8} \, Pa = 3.9_5 \, mol \, h^{-1}$$

$$r_{Soil-W}^{Runoff-Solids} = D_{Soil-W}^{Runoff-Solids} f_{Soil} = 3.5_8 \times 10^8 \, mol \, Pa^{-1} h^{-1} \times 4.0_7 \times 10^{-8} \, Pa = 14._6 \, mol \, h^{-1}$$

Total rates for intermedia transfer are obtained by the product of the source medium fugacity multiplied by the sum of all D-values for transport processes from the source to the destination medium.

Total air-to-water transfer:

$$\begin{aligned} r_{AW}^{T-A} &= \left(D_{AW}^{Diff-Ov} + D_{AW}^{Q} + D_{AW}^{Rain} \right) f_A \\ &= \left(2.01_3 \times 10^7 + 4.5_4 \times 10^8 + 2.16_0 \times 10^7 \right) mol \, Pa^{-1} h^{-1} \times 4.6_3 \times 10^{-7} \, Pa \\ &= 2.3_0 \times 10^2 \, mol \, h^{-1} \end{aligned}$$

Total water-to-air transfer:

$$r_{AW}^{T-W} = D_{AW}^{Diff-Ov} f_W = 2.01_3 \times 10^7 \, mol \, Pa^{-1} h^{-1} \times 1.7_8 \times 10^{-7} \, Pa = 3.5_9 \, mol \, h^{-1}$$

Total air-to-soil transfer:

$$\begin{aligned} r_{A-Soil}^{T-A} &= \left(D_{A-Soil}^{Diff-Ov} + D_{A-Soil}^{Q} + D_{A-Soil}^{Rain} \right) f_A \\ &= \left(1.81_5 \times 10^7 + 4.0_9 \times 10^9 + 1.94_4 \times 10^8 \right) mol \, Pa^{-1} h^{-1} \times 4.6_3 \times 10^{-7} \, Pa \\ &= 1.9_9 \times 10^3 \, mol \, h^{-1} \end{aligned}$$

Total soil-to-air transfer:

$$r_{A-Soil}^{T-Soil} = D_{A-Soil}^{Diff-Ov} f_{Soil} = 1.81_5 \times 10^7 \, mol\,Pa^{-1}h^{-1} \times 4.0_7 \times 10^{-8}\,Pa = 0.73_8 \, mol\,h^{-1}$$

Total sediment-to-water transfer:

$$\begin{aligned} r_{Sed-W}^{T-Sed} &= \left(D_{Sed-W}^{Diff-Ov} + D_{Sed-W}^{Resusp}\right) f_{Sed} \\ &= \left(2.16_0 \times 10^7 + 1.5_9 \times 10^9\right) mol\,Pa^{-1}h^{-1} \times 5.2_7 \times 10^{-7}\,Pa \\ &= 8.5_0 \times 10^2 \, mol\,h^{-1} \end{aligned}$$

Total water-to-sediment transfer:

$$\begin{aligned} r_{Sed-W}^{T-W} &= \left(D_{Sed-W}^{Diff-Ov} + D_{Sed-W}^{Dep}\right) f_W \\ &= \left(2.16_0 \times 10^7 + 1.2_4 \times 10^{10}\right) mol\,Pa^{-1}h^{-1} \times 1.7_8 \times 10^{-7}\,Pa \\ &= 2.2_2 \times 10^3 \, mol\,h^{-1} \end{aligned}$$

Total soil-to-water transfer:

$$\begin{aligned} r_{Soil-W}^{T-Soil} &= \left(D_{Soil-W}^{Runoff-W} + D_{Soil-W}^{Runoff-Solids}\right) f_{Soil} \\ &= \left(9.71_9 \times 10^7 + 3.5_8 \times 10^8\right) mol\,Pa^{-1}h^{-1} \times 4.0_7 \times 10^{-8}\,Pa \\ &= 18._5 \, mol\,h^{-1} \end{aligned}$$

We can also calculate net rates of transfer between media by taking the difference between total rates of transfer for pairs of media:

Net air-to-water transfer:

$$r_{AW}^{Net} = r_{AW}^{T-A} - r_{AW}^{T-W} = 2.3_0 \times 10^2 \, mol\,h^{-1} - 3.5_9 \, mol\,h^{-1} = 2.2_6 \times 10^2 \, mol\,h^{-1}$$

Net air-to-soil transfer:

$$r_{A-Soil}^{Net} = r_{A-Soil}^{T-A} - r_{A-Soil}^{T-Soil} = 1.9_9 \times 10^3 \, mol\,h^{-1} - 0.73_8 = 1.9_9 \times 10^3 \, mol\,h^{-1}$$

Net sediment-to-water transfer:

$$r_{Sed-W}^{Net} = r_{Sed-W}^{T-Sed} - r_{Sed-W}^{T-SedW} = 8.5_0 \times 10^2 \, mol\,h^{-1} - 2.2_2 \times 10^3 \, mol\,h^{-1} = -1.3_7 \times 10^3 \, mol\,h^{-1}$$

Net soil-to-water transfer:

$$r_{Soil-W}^{Net} = r_{Soil-W}^{T-Soil} - r_{Soil-W}^{T-W} = 18._5 \, mol\,h^{-1} - 0 = 18._5 \, mol\,h^{-1}$$

All such calculated rates of transfer give insight into the main pathways by which the chemical moves between media.

Screenshots of the results from this Level III calculation are given in Figures 5.8–5.10.

Level III Results

Chemical Name	Benzo[a]pyrene-Hughes et al. 2012
Environment Name	EQC-2012-Hughes et al.

Mass Balance

	Emission Rate	
	kg/h	mol/h
Air	1000.0	3963.3
Water	1000.0	3963.3
Soil	1000.0	3963.3
Sediment	0.0	0.0
	Total:	11890

	Inflow Rate		Concentration			
	kg/h	mol/h	ng/m³	ng/L	kg/m³	mol/m³
Air	0.0	0.0	0.000E+00	-	0.000E+00	0.000E+00
Water	0.0	0.0	-	0.000E+00	0.000E+00	0.000E+00
	Total:	0.0				
Total	3000.0	11889.9				

	kg	mol
Total Amount of Chemical in System	4.768E+07	1.890E+08

	Loss rate		D Value	Res. Time	
	kg/h	mol/h	m³/h	hours	days
Advection	8.409E+02	3.333E+03	1.944E+10	56699.77	2362.49
Reaction	2.159E+03	8.557E+03	1.670E+11	22084.48	920.19
Overall	3.000E+03	1.189E+04	1.864E+11	15893.85	662.24

Fugacity

	Fugacity	VZ
	Pa	mol/Pa
Air	4.6344E-07	2.674E+11
Water	1.78E-07	7.043E+12
Soil	4.07E-08	3.581E+15
Sediment	5.27E-07	7.958E+13
	Total:	3.668E+15
Fugacity Ratio	3.284E-02	
Sub-cooled Vapour Pressure	2.131E-05	

FIGURE 5.8 Screenshot of the Mass Balance and Fugacity panels from the Results page of the Level III model available from the CEMC website, with output for benzo[a]pyrene.

Phase Properties

	Z Value	Concentration			Quantity		
	mol/m³*Pa	mol/m³	g/m³	µg/g	mol	kg	%
Air							
Bulk	2.67E-03	1.239E-09	3.127E-07	2.606E-04	1.239E+05	3.127E+04	0.1
Air Vapour	4.0340E-04	1.869E-10	4.717E-08	3.931E-05	1.869E+04	4.717E+03	0.0
Aerosol	1.14E+08	52.63	1.328E+04	6.639E+03	1.053E+05	2.656E+04	0.1
Water							
Bulk	35.2	6.272E-06	1.582E-03	1.582E-03	1.254E+06	3.165E+05	0.7
Water	21.60	3.847E-06	9.706E-04	9.706E-01	7.693E+05	1.941E+05	0.4
Susp. Particles	2.49E+06	4.429E-01	1.117E+02	7.449E+01	4.429E+05	1.117E+05	0.2
Fish	1.18E+06	2.109E-01	5.321E+01	5.321E+01	4.218E+04	1.064E+04	0.0
Soil							
Bulk	1.99E+05	8.091E-03	2.042E+00	1.361E+00	1.456E+08	3.675E+07	77.1
Air	4.03E-04	1.641E-11	4.140E-09	3.450E-06	5.907E-02	1.490E-02	0.0
Water	21.60	8.785E-07	2.216E-04	2.216E-04	4.744E+03	1.197E+03	0.0
Solids	3.98E+05	1.618E-02	4.083E+00	1.701E+00	1.456E+08	3.675E+07	77.1
Sediment							
Bulk	1.59E+05	8.391E-02	2.117E+01	1.654E+01	4.196E+07	1.059E+07	22.2
Water	21.60	1.139E-05	2.873E-03	2.873E-03	4.555E+03	1.149E+03	0.0
Solids	7.96E+05	4.195E-01	1.058E+02	4.410E+01	4.195E+07	1.058E+07	22.2

Advection

	Res. Time		Flow Rate	D Value	Removal Rate		% of Total
	hours	days	m³/h	mol/Pa*h	kg/h	mol/h	Losses
Air							
Bulk	100.00	4.17	1.000E+12	2.67E+09	312.7377552	1.24E+03	10.4
Air Vapour	-	-	1.000E+12	4.03E+08	4.717E+01	1.869E+02	1.6
Aerosol	-	-	2.000E+01	2.27E+09	2.656E+02	1.053E+03	8.9
Water							
Bulk	1000.00	41.67	2.000E+08	7.04E+09	3.165E+02	1.254E+03	10.5
Water	-	-	2.000E+08	4.32E+09	1.941E+02	7.693E+02	6.5
Susp. Particles	-	-	1.000E+03	2.49E+09	1.117E+02	4.429E+02	3.7
Fish	-	-	2.000E+02	2.37E+08	1.064E+01	4.218E+01	0.4
Soil							
Bulk	1.00E+11	4.17E+09	-	-	-	-	-
Sediment							
Bulk	5.00E+04	2.08E+03	1.000E+04	1.59E+09	2.117E+02	8.391E+02	7.1
Total				1.13E+10	8.409E+02	3.333E+03	28.03

Reaction

	Half-Life		Rate Constant	D Value	Removal Rate		% of Total
	hours	days	1/h	mol/Pa*h	kg/h	mol/h	Losses
Air							
Bulk	-	-	-	1.09E+09	1.275E+02	5.053E+02	4.25E+00
Air Vapour	170.0	7.08E+00	4.076E-03	1.64E+08	1.923E+01	7.621E+01	6.41E-01
Aerosol	1.70E+02	7.08E+00	4.076E-03	9.26E+08	1.083E+02	4.291E+02	3.61E+00
Water							
Bulk	-	-	-	8.91E+09	4.002E+02	1.586E+03	1.33E+01
Water	1700.00	70.83	4.076E-04	1.76E+09	7.913E+01	3.136E+02	2.64E+00
Susp. Particles	1.70E+03	7.08E+01	4.076E-04	1.01E+09	4.555E+01	1.805E+02	1.52E+00
Fish	2.68E+01	1.12E+00	2.589E-02	6.13E+09	2.755E+02	1.092E+03	9.18E+00
Soil							
Bulk	17000.00	708.33	4.076E-05	1.46E+11	1.498E+03	5.937E+03	4.99E+01
Sediment							
Bulk	55000.00	2291.67	1.260E-05	1.003E+09	1.334E+02	5.286E+02	4.45E+00
Total				1.57E+11	2.159E+03	8.557E+03	71.97

FIGURE 5.9 Screenshot of the Phase Properties, Advection and Reaction panels from the Results page of the Level III model available from the CEMC website, with output for benzo[a]pyrene.

Intermedia Transport

	Half Times		Equiv. Flows	D Value	Transport Rate	
	hours	days	m³/h	mol/Pa*h	kg/h	mol/h
Air to water	3.74E+02	1.56E+01	1.854E+11	4.96E+08	57.99	229.84333
Air to soil	4.31E+01	1.80E+00	1.608E+12	4.30E+09	502.87	1993.03
Water to air	2.42E+05	1.01E+04	5.717E+05	2.01E+07	0.90	3.59
Water to sediment	3.92E+02	1.63E+01	3.537E+08	1.25E+10	559.67	2218.13
Soil to air	1.37E+08	5.70E+06	9.123E+01	1.81E+07	0.19	0.738
Soil to water	1362970.78	56790.45	2.289E+03	4.55E+08	4.67E+00	1.85E+01
Sediment to water	34197.01	1424.88	1.013E+04	1.61E+09	2.15E+02	8.50E+02

Process D Values

All D values are in units of mol/Pa*h

Air-water diffusion (air-side)	2.017E+07
Air-water diffusion (water-side)	1.080E+10
Air-water diffusion (total)	2.013E+07

Rain dissolution to water	2.160E+07
Rain dissolution to soil	1.944E+08

Aerosol deposition to water (dry)	1.36E-02
Aerosol deposition to water (wet)	4.54E+08
Aerosol deposition to water (total)	4.54E+08

Aerosol deposition to soil (dry)	1.23E-01
Aerosol deposition to soil (wet)	4.09E+09
Aerosol deposition to soil (total)	4.09E+09

Soil to water runoff (water)	9.719E+07
Soil to water runoff (solids)	3.581E+08

Soil-air diffusion (air-phase)	7.26E+05
Soil-air diffusion (water-phase)	1.94E+07
Soil-air diffusion (boundary layer)	1.82E+08
Soil-air diffusion (total)	1.815E+07

Water-sediment diffusion	2.16E+07

Water to sediment deposition	1.24E+10
Sediment to water resuspension	1.59E+09

FIGURE 5.10 Screenshot of the Intercompartment Transport and Process D Values panels from the Results page of the Level III model available from the CEMC website, with output for benzo[a]pyrene.

6 Time-Variant Systems
Differential Equations

In this chapter, time-variant systems are introduced and developed. The equations for fugacity change with time are related to fates and individual fugacities in non-equilibrated media or compartments. Various time-dependent processes such as liquid evaporation and depuration flow are developed and discussed. Half-times for time-dependent change are introduced. The approach to solving time-dependent differential equations using Euler's method of numerical integration is demonstrated with Excel spreadsheets. The resulting time dependence of fugacity and concentration is shown for several scenarios such as simple degradation, sediment depuration, sediment burial, effects, and degradation influences. Losses from a river sediment due to depuration, burial and degradation are compared. The attainment of equilibrium by two systems reaching equifugacity is demonstrated for two addition scenarios. Finally, the time dependence of a four-compartment system of air, water, soil, and sediment is developed with release to each of these in sequence to show the different impacts on the resulting chemical distributions at steady state.

6.1 TIME DEPENDENCE OF FUGACITY IN UNSTEADY-STATE SYSTEMS

To this point, we have considered systems at steady state, for which concentrations, and therefore fugacities, in the various compartments or media do not change with time. Of course, most real environmental systems of interest are in a state of change, either gaining or losing some chemical or another in various compartments as time progresses. Dealing with such systems involves a somewhat different approach from what we have used so far, although many of the fugacity-based terms and expressions remain unchanged in this context.

The basic approach for working with time-variant systems is to write out differential expressions that represent the net rate of change in the molar amount of chemical in a given compartment, calculated as the difference between the input and output rates:

$$\frac{dm_i}{dt} = \sum r_i^{Inf} - \sum r_i^{Outf}$$

We want to have an expression for the rate of change in fugacity with time, so we need to relate fugacity to the total amount of matter in a compartment. This relationship is:

$$m_i = V_i C_i = V_i Z_i f_i$$

Therefore, we can substitute the molar amount in the compartment with this expression involving fugacity:

$$\frac{d(V_i Z_i f_i)}{dt} = \sum r_i^{Inf} - \sum r_i^{Outf}$$

Given that the volume and fugacity capacity of a given compartment are generally constant in environmental systems, we can isolate the rate of change in the fugacity as follows:

$$\boxed{\frac{df_i}{dt} = \left(\frac{1}{V_i Z_i}\right)\left[\sum r_i^{Inf} - \sum r_i^{Outf}\right]}$$

DOI: 10.1201/9781003657170-6

This is the general format of an expression for the rate of change in fugacity with time. The various rates of input and output are expressed as the products of D-values and the appropriate fugacity for the compartment *from which* the transfer originates. For processes involving input to the compartment in question, the associated fugacity will be that of the source compartment. For output processes, the fugacity used is that of the compartment from which the transfer originates.

6.2 TIME DEPENDENCE OF DEPURATION AND EVAPORATION

The simplest time-dependent scenario involves an environmental compartment with an initial chemical fugacity transferring that chemical to another compartment that is rapidly refreshed so that it maintains a negligible concentration. Such a scenario might be depuration from an initially contaminated compartment, or vaporization transfer of a chemical from water to air. Considering the latter, the rate of transfer from water to air will be determined by:

$$\frac{df_W}{dt} = \left(\frac{1}{V_W Z_W}\right)\left(D_{AW}^{Diff-Ov} f_A - D_{AW}^{Diff-Ov} f_W\right)$$

Figure 6.1 illustrates this scenario graphically.

Since the air compartment is assumed here to be constantly and efficiently refreshed with clean air, the contribution from air to water diffusive transfer is zero, and the rate law for water fugacity simplifies to:

$$\frac{df_W}{dt} = \left(\frac{1}{V_W Z_W}\right)\left(0 - D_{AW}^{Diff-Ov} f_W\right)$$

This differential rate law is easily integrated to generate a first-order decay expression. First separating variables, we have:

$$\frac{df_W}{f_W} = -\left(\frac{D_{AW}^{Diff-Ov}}{V_W Z_W}\right)dt$$

Integrating from time zero to an arbitrary time t, we have:

$$\int_{f_0}^{f_t} \frac{df_W}{f_W} = -\left(\frac{D_{AW}^{Diff-Ov}}{V_W Z_W}\right)\int_{t_0}^{t} dt$$

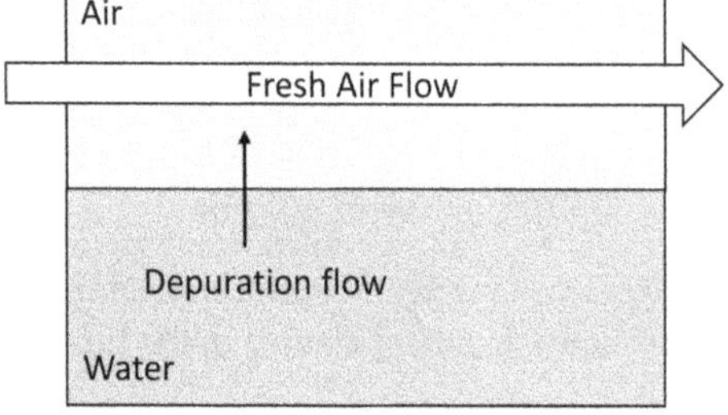

FIGURE 6.1 Depuration flow from water to air with a constant refreshing stream of air flowing over the water surface.

$$\ln(f_W)\Big|_{f_0}^{f_t} = -\left(\frac{D_{AW}^{Diff-Ov}}{V_W Z_W}\right) t\Big|_{t_0}^{t}$$

$$\ln(f_W^t) = \ln(f_W^{t_0}) - \left(\frac{D_{AW}^{Diff-Ov}}{V_W Z_W}\right) t$$

Taking antilogarithms, we obtain the first-order exponential decay expression:

$$f_W^t = f_W^{t_0} e^{-\left(\frac{D_{AW}^{Diff-Ov}}{V_W Z_W}\right)t}$$

Here, we can recognize that the rate constant for first-order decay is given by:

$$\boxed{k = \left(\frac{D_{AW}^{Diff-Ov}}{V_W Z_W}\right)}$$

The characteristic response time for the system is the reciprocal of this value, namely:

$$\boxed{\tau = \frac{1}{k} = \left(\frac{V_W Z_W}{D_{AW}^{Diff-Ov}}\right)}$$

This corresponds, in this application, to the time it takes for the fugacity or concentration to drop to exp (–1) = 0.368 of the original value, since:

$$\frac{f_W^\tau}{f_W^{t_0}} = e^{-\left(\frac{D_{AW}^{Diff-Ov}}{V_W Z_W}\right)\tau}$$

$$\ln\left(\frac{f_W^\tau}{f_W^{t_0}}\right) = -\left(\frac{D_{AW}^{Diff-Ov}}{V_W Z_W}\right)\tau = -\left(\frac{D_{AW}^{Diff-Ov}}{V_W Z_W}\right)\left(\frac{V_W Z_W}{D_{AW}^{Diff-Ov}}\right) = -1$$

$$\therefore \frac{f_W^\tau}{f_W^{t_0}} = e^{-1} = 0.368$$

The half-time for transfer is:

$$\tau_{1/2} = \frac{-\ln(0.5)}{k} = -\ln(0.5)\tau = -\ln(0.5)\left(\frac{V_W Z_W}{D_{AW}^{Diff-Ov}}\right)$$

The time to 95% transfer is:

$$\tau_{.95} = \frac{-\ln(0.05)}{k} = 3\tau = 3\left(\frac{V_W Z_W}{D_{AW}^{Diff-Ov}}\right)$$

Worked Example 6.1

A water body of volume $1.00 \times 10^4\ m^3$ is initially contaminated with a chemical at a concentration of $1.00 \times 10^{-4}\ mol\ m^{-3}$. The Henry's Law constant for this chemical is $50.0\ Pa\ m^3\ mol^{-1}$. Determine the decrease in fugacity after one day of evaporative transfer from water to air in a brisk breeze, across an air–water interfacial area of $1.00 \times 10^3\ m^2$. Use mass transfer coefficients of $50\ m\ h^{-1}$ and $0.50\ m\ h^{-1}$ for air and water, respectively, and a system temperature of 15.0°C. What are the characteristic and half-times for this transfer?

To proceed, we need the fugacity capacity of the chemical in air and in water. The air value is:

$$\boxed{Z_A = \frac{1}{RT}}$$

$$Z_A = \frac{1}{8.31446\,m^3 Pa\,K^{-1} mol^{-1} \times (15.0 + 273.15)K} = 4.174_0 \times 10^{-4} mol\,Pa^{-1} m^{-3}$$

The water fugacity capacity can be obtained from the Henry's Law constant:

$$\boxed{Z_W = \frac{1}{H}}$$

$$Z_W = \frac{1}{50.0\,Pa\,m^3\,mol^{-1}} = 2.00_0 \times 10^{-2} mol\,Pa^{-1} m^{-3}$$

The fugacity of the chemical in the water at the start of the transfer is given by:

$$f_W = \frac{C_W^{t=0}}{Z_W}$$

$$f_W = \frac{1.00 \times 10^{-4} mol\ m^{-3}}{2.00_0 \times 10^{-2} mol\,Pa^{-1} m^{-3}} = 5.00_0 \times 10^{-3} Pa$$

To determine the overall *D*-value for diffusive transfer between water and air, we need to know the individual interfacial *D*-values. These are obtained from the triple product of $k^M AZ$:

$$D_{AW}^{Diff-A} = k_{AW}^{M-A} A_{AW} Z_A$$

$$D_{AW}^{Diff-A} = 50.\ m\ h^{-1} \times 1.00 \times 10^3 m^2 \times 4.174_0 \times 10^{-4}\ mol\,Pa^{-1} m^{-3} = 2.0_9 \times 10^1 mol\,Pa^{-1} h^{-1}$$

$$D_{AW}^{Diff-W} = k_{AW}^{M-W} A_{AW} Z_W$$

$$D_{AW}^{Diff-W} = 5.0 \times 10^{-1}\ m\ h^{-1} \times 1.00 \times 10^3 m^2 \times 2.00_0 \times 10^{-2} mol\,Pa^{-1} m^{-3} = 1.0_0 \times 10^1 mol\,Pa^{-1} h^{-1}$$

The overall *D*-value for diffusive transfer is obtained from the reciprocal of the sum of the reciprocal interfacial *D*-values for air and water:

$$\boxed{D_{AW}^{Diff-Ov} = \frac{1}{\left(\frac{1}{D_{AW}^{Diff-A}} + \frac{1}{D_{AW}^{Diff-W}}\right)}}$$

$$D_{AW}^{Diff-Ov} = \frac{1}{\left(\frac{1}{2.0_9 \times 10^1 mol\,Pa^{-1} h^{-1}} + \frac{1}{1.0_0 \times 10^1 mol\,Pa^{-1} h^{-1}}\right)} = 6.7_6\,mol\,Pa^{-1} h^{-1}$$

The differential rate expression for the dependence of the water fugacity on time:

$$\frac{df_W}{dt} = \left(\frac{1}{V_W Z_W}\right)\left(0 - D_{AW}^{Diff-Ov} f_W\right)$$

This differential rate law integrates to:

$$f_W^t = f_W^{t_0} e^{\left(-\left[\frac{D_{AW}^{Diff-Ov}}{V_W Z_W}\right]t\right)}$$

The rate constant for this process is given by:

$$k = \left[\frac{D_{AW}^{Diff-Ov}}{V_W Z_W}\right] = \left(\frac{6.7_6\, mol\, Pa^{-1}h^{-1}}{1.00 \times 10^4 m^3 \times 2.00_0 \times 10^{-2} mol\, Pa^{-1}m^{-3}}\right) = 3.3_8 \times 10^{-2} h^{-1}$$

After 24 hours, the fugacity will have decreased to:

$$f_W^{t=24} = f_W^{t_0} e^{-kt}$$

$$f_W^{t=24} = 5.00_0 \times 10^{-3} Pa \times e^{\left(-3.3_8 \times 10^{-2} h^{-1} \times 24 h\right)}$$

$$f_W^{t=24} = 2.2_2 \times 10^{-3} Pa$$

The characteristic response time for the system is:

$$\tau = \frac{1}{k} = \frac{1}{3.3_8 \times 10^{-2} h^{-1}} = 29._6\, h$$

The half-time for transfer from water to air is:

$$\tau^{1/2} = -\ln(0.5)\tau = -\ln(0.5) \times 29._6\, h = 20._5\, h$$

6.3 SOLVING TIME-DEPENDENT PROBLEMS WITH NUMERICAL INTEGRATION

Although the example in Section 6.2 is relatively trivial, most realistic environmental modelling scenarios involve rather complicated expressions for the differential rate of change of fugacity in multiple compartments. Each compartment's rate law depends on the fugacities of one or more other compartments which are also changing with time. Such equations are a challenge to solve analytically. However, setting up a numerical solution to this type of problem is fairly straightforward with a spreadsheet. The simplest approach is "Euler's method" which involves setting initial values for the unknown fugacities based on some starting criteria, and then calculating and applying the rate of change in the various fugacities over a relatively small time duration (time step, Δt), using the finite difference approximation:

$$\frac{df_i}{dt} \approx \frac{\Delta f_i}{\Delta t}$$

In the limit of an infinitely small time step Δt, this approximation becomes an equality. Practically, it is only necessary that the value of $\frac{df_i}{dt}$ does not change appreciably over the chosen time step. A conservative estimate for the maximum magnitude of Δt that meets this criterion is given by 5% of the shortest response time for the compartments in question. The characteristic response time for any given compartment is defined as:

$$\tau_i = \frac{V_i Z_i}{D_i^L}$$

where D_i^L is the sum of all D-values for losses from compartment "i". Therefore, such a conservative estimate of the appropriate time step for Euler's method integration is given by:

$$\Delta t \le 0.05 \times \tau_i = \frac{0.05 \times V_i Z_i}{D_i^L}$$

Note that in this equation, "i" refers to the compartment with the shortest response time.

Once we have the appropriate time step in hand, the only spreadsheet setup question that remains is to determine for how long to run the simulation. In principle, the choice of time range is determined from the longest response time in the system, since that compartment will be the slowest to reach any steady-state value.

It is not practical to achieve 100% completion in any simulation involving first-order kinetics. Rather, a reasonable alternative which we employ here is the time to achieve 95% completion of loss from the compartment with the longest response time, for which time to 95% completion in the compartment in question ("i") will be:

$$\tau^{.95} = -\ln(0.05)\tau_i \simeq \frac{3 \times V_i Z_i}{D_i^L}$$

Word of Warning! One or more compartments may have extremely long response times, making the simulation both exceptionally long and compress the actual area of interest into a tiny fraction of the simulation plot. As well, the transformation of this slow-response compartment may be irrelevant to the point of the modelling exercise. In such a case, this compartment's response time should be ignored when determining the time range to be employed.

We shall begin by looking at the application of Euler's method of numerical integration for the relatively simple case of a single compartment whose fugacity and concentration are decreasing exponentially. This problem can be easily solved analytically, so numerical integration is unnecessary. However, it is an excellent starting point for getting comfortable with Euler's method, with a relatively transparent problem.

Worked Example 6.2

Use Euler's method to numerically integrate the differential rate equation from Worked Example 6.1 and plot the variation in water fugacity and concentration with time until 95% completion is reached.

The appropriate time step is conservatively given by:

$$\Delta t \leq 0.05 \times \tau_i = 0.05 \times 29._6\, h = 1.4_8\, h$$

This is very close to 1.5 h, so we will use that time step to make the presentation more agreeable.

The time range for the simulation is straightforward to determine with only one loss process from the water, for which the time to 95% completion is:

$$\tau^{.95} = \frac{3 \times V_W Z_W}{D_{AW}^{Diff-Ov}} = \frac{3 \times 1.00 \times 10^4 m^3 \times 2.00_0 \times 10^{-2} mol\, Pa^{-1} m^{-3}}{6.7_6\, mol\, Pa^{-1} h^{-1}} = 88._7 h$$

Given this time range, the number of simulation time steps needed is:

$$\frac{88._7 h}{1.5 h} \simeq 59$$

The initial fugacity is already calculated as $5.00_0 \times 10^{-3} Pa$. The differential rate law for change is given by $-kf_w$:

$$\frac{df_W}{dt} = -kf_W = -\left(\frac{D_{AW}^{Diff-Ov}}{V_W Z_W}\right)(f_W) = -3.3_8 \times 10^{-2} h^{-1}(f_W)$$

Since we are using numerical integration, it is not necessary to analytically integrate this rate law. We simply go ahead and use the differential rate law itself. Setting up the spreadsheet integration of this equation requires two columns, one for time and one for the water fugacity at that time. Since we also want the water concentration, we can add a third column for that, which will be calculated from $C = Zf$. We will need to make repeated reference to the time step, the fugacity capacity of water, and the rate constant for transfer, so these should be entered on the sheet as well. The first three rows so far, with the parameters and constants entered, and the calculation formula for the concentration at time, $t = 0$ revealed, could look like Figure 6.2.

Note the use of the "$" symbol before the column and row designation of the Z_W cell address, so that when we copy this line, it will remain referenced to that cell. The sheet when the concentration formula is not revealed is shown in Figure 6.3.

The next row is the first and only complicated row to set up, after which all other rows will follow by copy and paste. The third row holds the values for 1.5 *h* after the conditions of the second row, so the time will be 1.5 *h*. This is coded in the second time slot as "=A2+$F$2", as shown in Figure 6.4. This will look as shown in Figure 6.5 when the A3 cell equation is not revealed.

	A	B	C	D	E	F	G
1	Time /h	fW(t)/ Pa	C /mol m-3		Zw	2.0000E-02	mol Pa-1m-3
2	0	5.00000E-03	=B2*$F$1		DelT	1.5	h
3					k	3.38E-02	h-1

FIGURE 6.2 Screenshot of the first three rows of an Excel worksheet for Worked Example 6.2 showing the equation used to calculate concentration from fugacity and the fugacity capacity.

	A	B	C	D	E	F	G
1	Time /h	fW(t)/ Pa	C /mol m-3		Zw	2.0000E-02	mol Pa-1m-3
2	0	5.00000E-03	1.00E-04		DelT	1.5	h
3					k	3.38E-02	h-1

FIGURE 6.3 Screenshot of the first three rows of an Excel worksheet for Worked Example 6.2 showing the calculated concentration value from fugacity and the fugacity capacity.

	A	B	C	D	E	F	G
1	Time /h	fW(t)/ Pa	C /mol m-3		Zw	2.0000E-02	mol Pa-1m-3
2	0	5.00000E-03	1.00E-04		DelT	1.5	h
3	=A2+$F$2				k	3.38E-02	h-1

FIGURE 6.4 Screenshot of the first three rows of an Excel worksheet for Worked Example 6.2 showing the equation used to calculate the time after one time step from the previous row's time value and the time step magnitude value.

	A	B	C	D	E	F	G
1	Time /h	fW(t)/ Pa	C /mol m-3		Zw	2.0000E-02	mol Pa-1m-3
2	0	5.00000E-03	1.00E-04		DelT	1.5	h
3	1.5				k	3.38E-02	h-1

FIGURE 6.5 Screenshot of the first three rows of an Excel worksheet for Worked Example 6.2 showing the time after one time step from the previous row's time value and the time step magnitude value.

	A	B	C	D	E	F	G
1	Time /h	fW(t)/ Pa	C /mol m-3		Zw	2.0000E-02	mol Pa-1m-3
2	0	5.00000E-03	1.00E-04		DelT	1.5	h
3	1.5	=B2*(1-$F$3*$F$2)			k	3.38E-02	h-1

FIGURE 6.6 Screenshot of the first three rows of an Excel worksheet for Worked Example 6.2 showing the equation used to calculate the fugacity one time step from the previous row's fugacity, the rate constant and the time step magnitude value.

The fugacity at time $t = 1.5\ h$ is estimated in Euler's method by finite difference, such that:

$$f_W^{t+\Delta t} = f_W^t + \left(\frac{df_W}{dt}\right)\Delta t \simeq f_W^t + \left(-k \times f_W^t\right)\Delta t = f_W^t\left(1 - k\Delta t\right)$$

Therefore, the third row at column B (fugacity at $t = 1.5\ h$) is coded as "=B2*(1 – $F$3*$F$2)", as shown in Figure 6.6.

This will look like Figure 6.7 when the cell formula is not revealed.

Note that, as expected, the fugacity in the water has decreased by a small amount, about 5%.

The concentration can be calculated in the appropriate cell for time t as $C_W^t = Z_W f_W^t$ as shown in Figure 6.8 which looks like the following when the formula is not revealed (Figure 6.9).

Since concentration is proportional to fugacity, the decrease is also about 5%.

	A	B	C	D	E	F	G
1	Time /h	fW(t)/ Pa	C /mol m-3		Zw	2.0000E-02	mol Pa-1m-3
2	0	5.00000E-03	1.00E-04		DelT	1.5	h
3	1.5	4.74650E-03			k	3.38E-02	h-1

FIGURE 6.7 Screenshot of the first three rows of an Excel worksheet for Worked Example 6.2 showing the calculated fugacity one time step from the previous row's fugacity.

	A	B	C	D	E	F	G
1	Time /h	fW(t)/ Pa	C /mol m-3		Zw	2.0000E-02	mol Pa-1m-3
2	0	5.00000E-03	1.00E-04		DelT	1.5	h
3	1.5	4.74650E-03	=$F$1*B3		k	3.38E-02	h-1

FIGURE 6.8 Screenshot of the first three rows of an Excel worksheet for Worked Example 6.2 showing the equation used to calculate the concentration after one time step from the fugacity and the fugacity capacity.

	A	B	C	D	E	F	G
1	Time /h	fW(t)/ Pa	C /mol m-3		Zw	2.0000E-02	mol Pa-1m-3
2	0	5.00000E-03	1.00E-04		DelT	1.5	h
3	1.5	4.74650E-03	9.49E-05		k	3.38E-02	h-1

FIGURE 6.9 Screenshot of the first three rows of an Excel worksheet for Worked Example 6.2 showing the concentration at time step 1.5 *h*.

Since we have taken the trouble to freeze the cell references to Z_W, k, and Δt, we can simply block the third-row entries A3 to C3 and drag the block down over enough cells to cover the required time, which is just under 89 *h*. Do this by clicking on the small square that appears in the bottom right-hand corner of the block once it's selected.

Word of Warning! To do this operation, you may need to choose "Enable fill handle and cell drag-and-drop" from the "Advanced" page of Excel Options for your sheet.

The first and last few cells over this range should be as shown in Figure 6.10a and b.

You can see that the final fugacity and concentration are just under 5.0% of the starting value, in accordance with our calculated number of required timesteps to reach this level of evaporative loss.

The progression of the fugacity or concentration change with time may now be plotted. The plot will appear as a marker-free line if you turn off "Markers" and turn on "Line" within the "Format Data Series" option of Excel, accessed by double-clicking on any of the data points. This will give a smooth line, and should look like Figure 6.11.

(a)

	A	B	C
1	Time /h	fW(t)/ Pa	C /mol m-3
2	0	5.00000E-03	1.00E-04
3	1.5	4.74650E-03	9.49E-05
4	3	4.50585E-03	9.01E-05
5	4.5	4.27741E-03	8.55E-05
6	6	4.06054E-03	8.12E-05

(b)

55	79.5	3.17216E-04	6.34E-06
56	81	3.01133E-04	6.02E-06
57	82.5	2.85865E-04	5.72E-06
58	84	2.71372E-04	5.43E-06
59	85.5	2.57613E-04	5.15E-06
60	87	2.44552E-04	4.89E-06

FIGURE 6.10 (a) and (b) Screenshot of the first six rows of an Excel worksheet for Worked Example 6.2 showing the calculated time, fugacity and concentration at each time step, followed by the same for time 79.5–87 *h*.

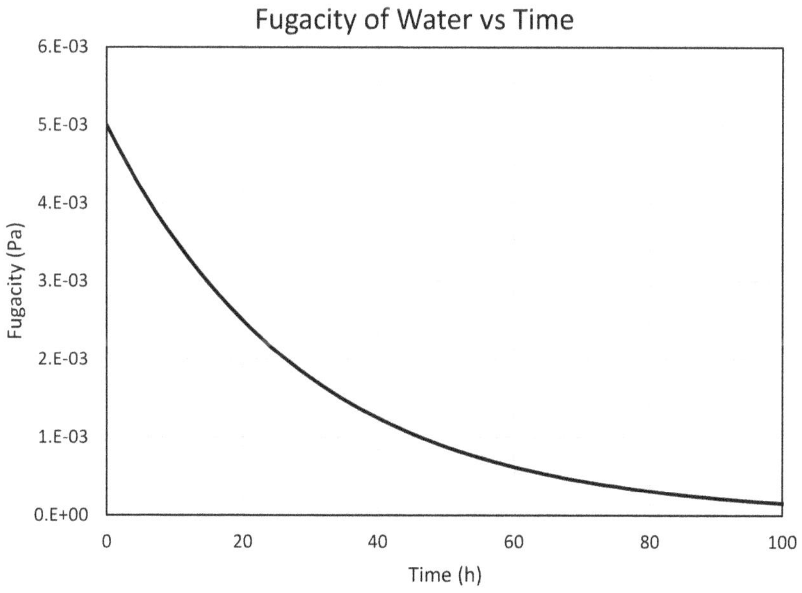

FIGURE 6.11 Variation in the fugacity of water as a function of time for a body of water with wind blowing over it.

6.4 SEDIMENT DEPURATION IN A RIVER

A scenario that is analogous to diffusive loss from water to a fast-moving air stream is depuration of a chemical from contaminated sediment in a flowing river. Here, for simplicity, we ignore all transformational losses in this example, and assume the sediment is static, with no deposition, resuspension or burial. In this admittedly unrealistic scenario, the only significant mechanism for sediment depuration is diffusive transfer from sediment to water. Figure 6.12 illustrates this scenario, which is completely analogous to air flow over water.

Worked Example 6.3

Consider sediment contaminated with a chemical, above which a sediment-free river is flowing, for which the environmental parameters given below apply. Assume there is no decomposition in the sediment, the chemical is not present in the upstream river, and the sediment layer does not undergo any changes due to deposition, resuspension, or burial. Calculate the appropriate time step for stable numerical integration of the loss kinetics, as well as the time required for 50% and 95% reduction in the concentration due to diffusive transfer of the chemical to the river water. Use the data below for a DDT-like chemical.

Chemical properties of a DDT-like chemical:

Henry's Law constant ($Pa\ m^3\ mol^{-1}$)	2.29
log K_{OW}	6.10
K_{MM-W} ($L\ kg^{-1}$)	1.00

Environmental parameters:

Water-sediment interfacial area (m^2)	1000
Water depth (m)	20.
Sediment depth (m)	0.030
Volume fraction of water in sediment, v_{Sed}^{f-W}	0.80
Mass fraction of organic carbon in organic matter, m_{OM}^{f-OC}	0.56
Molecular diffusivity of chemical in water B_{Sed}^{W} ($m^2\ h^{-1}$)	2.00×10^{-6}
Density of dry sediment organic matter, ρ_{OM} ($kg\,m^{-3}$)	1000
Density of dry sediment mineral matter, ρ_{MM} ($kg\,m^{-3}$)	2500
Mass fraction of organic carbon in sediment, m_{Sed}^{f-OC}	0.050
Sediment diffusion path length Y_{Sed} (m)	0.015

As always, we begin by calculating the fugacity capacity of the river water and the sediment. For the water, we have the Henry's Law constant, from which the fugacity capacity follows immediately:

$$Z_W = \frac{1}{H}$$

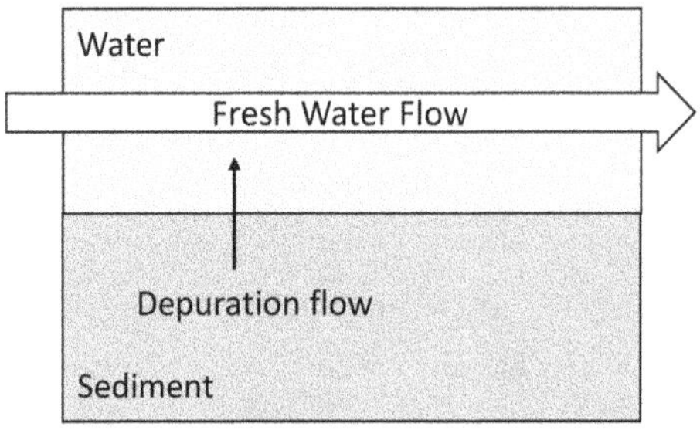

FIGURE 6.12 Schematic of sediment depuration to water that is constantly refreshed.

$$Z_W = \frac{1}{2.29\,Pa\,m^3 mol^{-1}}$$

$$Z_W = 4.36_7 \times 10^{-1} mol\,Pa^{-1}m^{-3}$$

The bulk sediment fugacity capacity must be calculated in the same manner as developed earlier. First, we determine the organic matter and mineral matter fugacity capacities:

$$Z_{OM} = m_{OM}^{f-OC} \times 0.35\,L\;kg^{-1} \times K_{OW} \times Z_W \times \left(\frac{\rho_{OM}\left(kg\,m^{-3}\right)}{1000\,L\,m^{-3}}\right)$$

$$Z_{OM} = 0.56 \times 0.35\,L\;kg^{-1} \times 10^{6.10} \times 4.36_7 \times 10^{-1} mol\,Pa^{-1}m^{-3} \times \left(\frac{1000\,kg\,m^{-3}}{1000\,L\,m^{-3}}\right)$$

$$Z_{OM} = 1.0_8 \times 10^5 mol\,Pa^{-1}m^{-3}$$

$$Z_{MM} = K_{MM-W}\left(L\,kg^{-1}\right) \times Z_W \times \left(\frac{\rho_{MM}\left(kg\,m^{-3}\right)}{1000\,L\,m^{-3}}\right)$$

$$Z_{MM} = 1.00\,L\,kg^{-1} \times 4.36_7 \times 10^{-1} mol\,Pa^{-1}m^{-3} \times \left(\frac{2500\,kg\,m^{-3}}{1000\,L\,m^{-3}}\right)$$

$$Z_{MM} = 1.09_2\,mol\,Pa^{-1}m^{-3}$$

The volume fraction of the water is given at 0.80. As in Chapter 2, the volume fraction of the solid matter in the sediment is deduced by difference:

$$v_{Sed}^{f-W} = 0.80$$

$$v_{Sed}^{f-Dry\,Sed} = 1 - v_{Sed}^{f-W} = 1 - 0.80 = 0.20$$

The organic carbon content of the dry sediment is provided as a mass fraction. Using the mass fraction of 0.56 for organic carbon content in organic matter (*OM*), the mass fraction of organic matter in the sediment is:

$$m_{Sed}^{f-OM} = \frac{m_{Sed}^{f-OC}}{m_{OM}^{f-OC}} = \frac{0.050}{0.56} = 0.089_3$$

The remaining mass fraction in the dry sediment must be mineral matter (*MM*), so that:

$$m_{Sed}^{f-MM} = m_{Sed}^{f-Dry\,Sed} - m_{Sed}^{f-OM} = 1 - 0.089_3 = 0.910_7$$

To calculate the bulk fugacity capacity for the overall sediment, we need to know the volume fractions of organic matter and mineral matter. As before, we can relate the volume and mass fractions of these as:

$$v_{Dry\,Sed}^{f-OM} = \left[\frac{1}{1+\left(\frac{m_{Sed}^{f-MM}/\rho_{MM}}{m_{Sed}^{f-OM}/\rho_{OM}}\right)}\right] = \left[\frac{1}{1+\left(\frac{0.910_7/2500\,kg\,m^{-3}}{0.089_3/1000\,kg\,m^{-3}}\right)}\right] = 0.19_7$$

$$v_{Dry\,Sed}^{f-MM} = \left[\frac{1}{1+\left(\frac{m_{Sed}^{f-OM}/\rho_{OM}}{m_{Sed}^{f-MM}/\rho_{MM}}\right)}\right] = \left[\frac{1}{1+\left(\frac{0.089_3/1000\,kg\,m^{-3}}{0.910_7/2500\,kg\,m^{-3}}\right)}\right] = 0.803_1$$

Finally, recognizing that the dry sediment is only 0.20 of the overall wet sediment by volume, we have the volume fractions for organic matter and mineral matter in the wet sediment as:

$$v_{Sed}^{f-OM} = 0.20 \times v_{Dry\,Sed}^{f-OM} = 0.20 \times 0.19_7 = 3.9_4 \times 10^{-2}$$

$$v_{Sed}^{f-MM} = 0.20 \times v_{Dry\,Sed}^{f-MM} = 0.20 \times 0.803_1 = 0.16_1$$

The bulk fugacity capacity can now be calculated:

$$Z_{Sed}^{Bulk} = v_{Sed}^{f-OM} Z_{OM} + v_{Sed}^{f-MM} Z_{MM} + v_{Sed}^{f-W} Z_W$$

$$Z_{Sed}^{Bulk} = 3.9_4 \times 10^{-2} \times 1.0_8 \times 10^5 mol\,Pa^{-1}m^{-3} + 0.16_1 \times 1.09_2\,mol\,Pa^{-1}m^{-3} + 0.80 \times 4.36_7 \times 10^{-1} mol\,Pa^{-1}m^{-3}$$

$$Z_{Sed}^{Bulk} = 4.2_4 \times 10^3 mol\,Pa^{-1}m^{-3}$$

The only loss process that we need to consider is diffusive transfer between sediment and water. We can calculate this D-value from the stated molecular diffusivity, using the following relation from sediment diffusion theory, which is similar to that used for soils in Chapter 4, but includes an additional correction term for void volume due to the sediment water content:

$$D_{Sed-W}^{Diff-Ov} = \left(\frac{B_{Sed-W}^{W} A_{Sed-W} Z_W}{Y_{Sed}} \right) \left(v_{Sed}^{f-W} \right)^{1.5}$$

$$D_{Sed-W}^{Diff-Ov} = \left(\frac{2.00 \times 10^{-6} m^2\,h^{-1} \times 1000\,m^2 \times 4.36_7 \times 10^{-1} mol\,Pa^{-1}m^{-3}}{1.5 \times 10^{-2} m} \right) 0.80^{1.5}$$

$$D_{Sed-W}^{Diff-Ov} = 4.1_7 \times 10^{-2} mol\,Pa^{-1}h^{-1}$$

The maximum time step can be calculated from this value, the sediment volume, and the sediment Z-value:

$$\Delta t \le 0.05 \times \tau_{Sed} = \frac{0.05 \times V_{Sed} Z_{Sed}^{Bulk}}{D_{Sed}^{L}}$$

$$\Delta t \le \frac{0.05 \times (1000\,m^2 \times 0.030\,m) \times 4.2_4 \times 10^3 mol\,Pa^{-1}m^{-3}}{4.1_7 \times 10^{-2} mol\,Pa^{-1}h^{-1}} = 1.5_3 \times 10^5 h$$

Here, we note that the time step for this DDT-like chemical corresponds to over 17.4 years! It is already clear that sediment remediation by diffusive transfer to a flowing river alone is a very slow and ineffective process and will take an extremely long time to achieve any significant level of reduction.

The half times are given by:

$$\tau_{Sed}^{1/2} = -\ln(0.5)\tau_{Sed} \simeq -\ln(0.5)\left(\frac{V_{Sed} Z_{Sed}^{Bulk}}{D_{Sed}^{L}} \right)$$

$$= \frac{-\ln(0.5) \times (1000\,m^2 \times 0.030\,m) \times 4.2_4 \times 10^3 mol\,Pa^{-1}m^{-3}}{4.1_7 \times 10^{-2} mol\,Pa^{-1}h^{-1}} = 2.1_2 \times 10^6 h$$

This half time corresponds to about 240 years.

For 95% reduction, we have:

$$\tau_{Sed}^{95\%} = -\ln(0.05)\tau_{Sed} \simeq \frac{3 \times V_{Sed} Z_{Sed}^{Bulk}}{D_{Sed}^{L}} = \frac{3 \times (1000\,m^2 \times 0.030\,m) \times 4.2_4 \times 10^3 mol\,Pa^{-1}m^{-3}}{4.1_7 \times 10^{-2} mol\,Pa^{-1}h^{-1}} = 9.1_7 \times 10^6 h$$

This result corresponds to 1046 years for the chemical to essentially clear from this sediment scenario. Clearly, remediation by "doing nothing" is not an effective strategy for this chemical if reasonable time frames are required, such as a typical human lifetime of < 10^2 years!

6.5 SEDIMENT DIFFUSIVE DEPURATION WITH SEDIMENT BURIAL

The convenience of using the fugacity approach is evident when one wishes to consider the impact of various parallel processes on the overall rate of change in a dynamic system. The following example demonstrates the ease with which additional processes may be incorporated into the overall model, because of the directly additive nature of *D*-values. In this next Worked Example, we consider the impact of adding an additional sediment loss process, namely burial, as illustrated in Figure 6.13.

Worked Example 6.4

Repeat the calculation of the time step for stable numerical integration, and the time for reduction of the concentration by 50% and 95% due to transfer from the sediment to the river water for the chemical in Worked Example 6.3. In addition to the data given and calculated there, incorporate a burial flux of 1.5 $g\,m^{-2}$ day^{-1}.

First, we must calculate the *D*-value for the burial process. The calculation first requires that we calculate the rate of sediment burial in $m^3\,h^{-1}$, which itself requires the density of the dry sediment. The dry sediment density is the volume-fraction-weighted sum of the organic and mineral content densities in the sediment:

$$\rho_{Sed}^{Dry} = v_{Dry\ Sed}^{f-OM} \times \rho_{Sed}^{OM} + v_{Dry\ Sed}^{f-MM} \times \rho_{Sed}^{MM}$$

$$= 0.19_7 \times 1000\,kg\,m^{-3} + 0.803_1 \times 2500\,kg\,m^{-3}$$

$$\rho_{Sed}^{Dry} = 2.20_5 \times 10^3\,kg\,m^{-3}$$

The burial rate follows as:

$$G_{Sed}^{Burial} = \frac{L_{Sed}^{Burial} \times A_{Sed-W}}{\rho_{Sed}^{Dry}} \times \frac{10^{-3}\,kg\,g^{-1}}{24\,h\,day^{-1}}$$

$$G_{Sed}^{Burial} = \frac{1.5\,g\,m^{-2}day^{-1} \times 1000.\,m^2}{2.20_5 \times 10^3\,kg\,m^{-3}} \times \frac{10^{-3}\,kg\,g^{-1}}{24\,h\,day^{-1}} = 2.8_3 \times 10^{-5}\,m^3\,h^{-1}$$

Now, the burial is assumed to carry no water but only remove solids from the sediment active layer. For this reason, the appropriate *Z*-value is that of the sediment solids only, excluding the

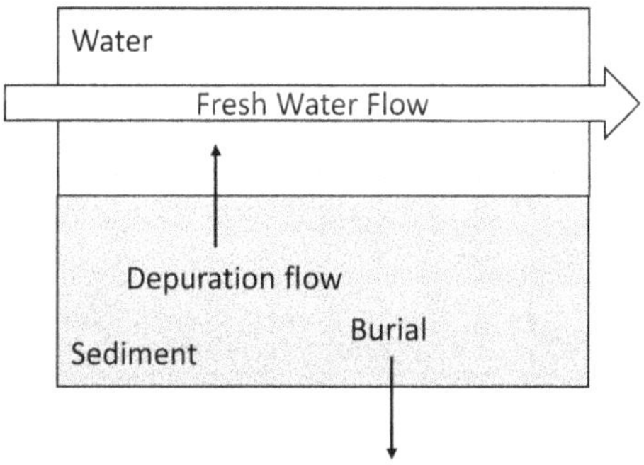

FIGURE 6.13 Schematic of sediment depuration to water that is constantly refreshed with sediment burial contributing to overall sediment losses.

water. This is calculated using the volume fractions of organic and mineral matter in the sediment solids and their respective Z-values:

$$Z_{DrySed}^{Solids} = v_{Dry\,Sed}^{f-OM} Z_{OM} + v_{Dry\,Sed}^{f-MM} Z_{MM}$$

$$Z_{DrySed}^{Solids} = 0.19_7 \times 1.0_8 \times 10^5 mol\,Pa^{-1}m^{-3} + 0.803_1 \times 1.09_2\,mol\,Pa^{-1}m^{-3}$$

$$Z_{DrySed}^{Solids} = 2.1_2 \times 10^4 mol\,Pa^{-1}m^{-3}$$

With this fugacity capacity in hand, the D-value for burial is:

$$D_{Sed}^{Burial} = G_{Sed}^{Burial} Z_{DrySed}^{Solids}$$

$$D_{Sed}^{Burial} = 2.8_3 \times 10^{-5} m^3\,h^{-1} \times 2.1_2 \times 10^4 mol\,Pa^{-1}m^{-3} = 6.0_1 \times 10^{-1} mol\,Pa^{-1}h^{-1}$$

The only change in the calculation of the time step, half time and 95% consumption rate is the recalculation of the D-value for total loss, in which we now include the burial process loss D-value in the sum:

$$D_{Sed}^{L} = D_{Sed-W}^{Diff-Ov} + D_{Sed}^{Burial}$$

$$D_{Sed}^{L} = 4.1_7 \times 10^{-2}\,mol\,Pa^{-1}h^{-1} + 6.0_1 \times 10^{-1} mol\,Pa^{-1}h^{-1} = 6.4_3 \times 10^{-1}\,mol\,Pa^{-1}h^{-1}$$

The recommended maximum step value can now be recalculated:

$$\Delta t \le 0.05 \times \tau_{Sed} = \frac{0.05 \times V_{Sed} Z_{Sed}^{Bulk}}{D_{Sed}^{L}}$$

$$\Delta t \le \frac{0.05 \times (1000\,m^2 \times 0.030\,m) \times 4.2_4 \times 10^3 mol\,Pa^{-1}m^{-3}}{6.4_3 \times 10^{-1} mol\,Pa^{-1}h^{-1}} = 9.9_0 \times 10^3 h$$

The maximum time step for the DDT-like chemical is now significantly reduced to a little over a year, due to the impact of the relatively efficient burial process, which is much faster than diffusion for this chemical.

The half time is given by:

$$\tau_{Sed}^{1/2} = -\ln(0.5)\tau_{Sed} \simeq -\ln(0.5)\left(\frac{V_{Sed} Z_{Sed}^{Bulk}}{D_{Sed}^{L}}\right)$$

$$= \frac{-\ln(0.5) \times (1000\,m^2 \times 0.030\,m) \times 4.2_4 \times 10^3 mol\,Pa^{-1}m^{-3}}{6.4_3 \times 10^{-1} mol\,Pa^{-1}h^{-1}} = 1.3_7 \times 10^5 h$$

The half time is reduced from about 240 to 16 years. Here we immediately see the impact of a much higher D-value for burial (roughly 10× bigger) compared to loss by diffusion to water for the DDT-like chemical.

For 95% reduction, we have:

$$\tau_{Sed}^{.95} = -\ln(0.05)\tau_{Sed} \simeq \frac{3 \times V_{Sed} Z_{Sed}^{Bulk}}{D_{Sed}^{L}} = \frac{3 \times (1000\,m^2 \times 0.030\,m) \times 4.2_4 \times 10^3 mol\,Pa^{-1}m^{-3}}{6.4_3 \times 10^{-1} mol\,Pa^{-1}h^{-1}} = 5.9_4 \times 10^5 h$$

These results correspond to a 95% reduction time of about 68 years. Burial has greatly improved the remediation prospects for this chemical (assuming the buried sediment stays buried!).

6.6 SEDIMENT DIFFUSIVE DEPURATION WITH SEDIMENT BURIAL AND DEGRADATION

As a final exercise with sediments, we can include degradation within the sediment, as the next Worked Example demonstrates. To do so, we will need to calculate the D-value from the half-time for degradation as follows:

$$D_{Sed}^{Deg} = k_{Sed}^{Deg} V_{Sed} Z_{Sed}^{Bulk}$$

Figure 6.14 illustrates this scenario.

Worked Example 6.5

For the chemical in Worked Example 6.4, recalculate the time step for stable numerical integration, the time for reduction of the concentration by 50% and 95% when including degradation within the sediment as a third loss mechanism. Use a degradation half time in sediment of $1.3 \times 10^5\ h$.

Given the degradation half time, we can determine the rate constant for first-order degradation as follows:

$$k_{Sed}^{Deg} = \frac{-\ln(0.5)}{\tau_{1/2}} = \frac{-\ln(0.5)}{1.3 \times 10^5 h} = 5.3_3 \times 10^{-6} h^{-1}$$

The D-value for degradation follows immediately:

$$\begin{aligned} D_{Sed}^{Deg} &= k_{Sed}^{Deg} \times V_{Sed} \times Z_{Sed}^{Bulk} \\ &= 5.3_3 \times 10^{-6} h^{-1} \times \left(1000\, m^2 \times 0.030\, m\right) \times 4.2_4 \times 10^3 mol\, Pa^{-1} m^{-3} \\ &= 6.7_9 \times 10^{-1} mol\, Pa^{-1} h^{-1} \end{aligned}$$

The maximum time step can now be recalculated once again, with the updated sum of loss D-values:

$$D_{Sed}^{L} = D_{Sed-W}^{Diff-Ov} + D_{Sed}^{Burial} + D_{Sed}^{Deg}$$

$$D_{Sed}^{L} = 4.1_7 \times 10^{-2} mol\, Pa^{-1} h^{-1} + 6.0_1 \times 10^{-1} mol\, Pa^{-1} h^{-1} + 6.7_9 \times 10^{-1} mol\, Pa^{-1} h^{-1}$$

$$D_{Sed}^{L} = 1.3_2\, mol\, Pa^{-1} h^{-1}$$

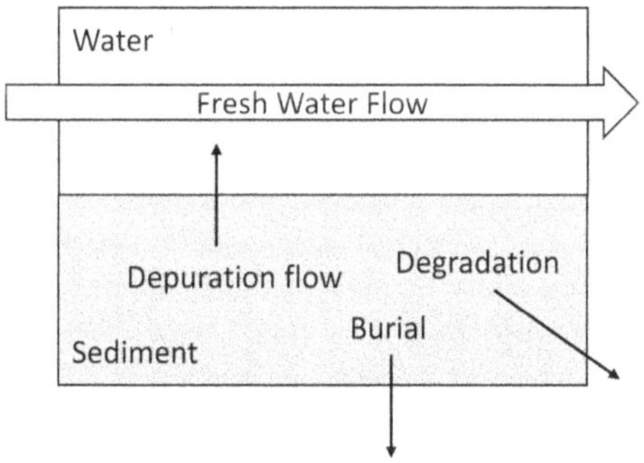

FIGURE 6.14 Schematic of sediment depuration to water that is constantly refreshed, and in which sediment burial and degradation contribute to sediment losses.

$$\Delta t \le 0.05 \times \tau_{Sed} = \frac{0.05 \times V_{Sed} Z_{Sed}^{Bulk}}{D_{Sed}^{L}}$$

$$\Delta t \le \frac{0.05 \times (1000\,m^2 \times 0.030\,m) \times 4.2_4 \times 10^3 mol\,Pa^{-1}m^{-3}}{1.3_2 mol\,Pa^{-1}h^{-1}} = 4.8_2 \times 10^3 h$$

The maximum time step for the DDT-like chemical is now further reduced to about half a year with degradation added in.

The half time is given by:

$$\tau_{Sed}^{1/2} = -\ln(0.5)\tau_{Sed} \approx -\ln(0.5)\left(\frac{V_{Sed} Z_{Sed}^{Bulk}}{D_{Sed}^{L}}\right)$$

$$= \frac{-\ln(0.5) \times (1000\,m^2 \times 0.030\,m) \times 4.2_4 \times 10^3 mol\,Pa^{-1}m^{-3}}{1.3_2 mol\,Pa^{-1}h^{-1}} = 6.6_8 \times 10^4 h$$

For 95% reduction, we have:

$$\tau_{Sed}^{95} = -\ln(0.05)\tau_{Sed} \approx \frac{3 \times V_{Sed} Z_{Sed}^{Bulk}}{D_{Sed}^{L}}$$

$$= \frac{3 \times (1000\,m^2 \times 0.030\,m) \times 4.2_4 \times 10^3 mol\,Pa^{-1}m^{-3}}{1.3_2\, mol\,Pa^{-1}h^{-1}}$$

$$= 2.8_9 \times 10^5 h$$

These results correspond to a half time of 7.6 years and time to 95% depletion of 33 years for this chemical. Degradation loss has further improved the remediation prospects of this chemical well beyond that of simple diffusive loss.

6.7 TIME DEPENDENCE OF SEDIMENT CHEMICAL LOSS

With all these results in hand we can compare the depuration under the circumstances of diffusion only, diffusion with burial, and diffusion with both burial and degradation for this chemical.

Worked Example 6.6

Plot the time dependence of the fugacity of the DDT-like chemical in the sediment for which you have determined D-values and maximum time steps for (i) diffusion only, (ii) diffusion with burial and (iii) diffusion, burial and degradation. Assume initial concentrations of the chemical in the water and sediment of $1.00 \times 10^{-4}\ g\,m^{-3}$ and $10.0\ \mu g\,g^{-1}$, respectively. Use a molar mass of $354.5\ g\,mol^{-1}$.

Before we can proceed, we need to put the concentration of the chemical into units of $mol\,m^{-3}$, using the appropriate conversions. It is straightforward for the water:

$$C_W^0 = \frac{1.00 \times 10^{-4} g\,m^{-3}}{354.5\,g\,mol^{-1}} = 2.82 \times 10^{-7} mol\,m^{-3}$$

To convert the sediment concentration from units of micrograms of chemical per gram of dry sediment, we will need the density of dry sediment from Worked Example 6.4, $\rho_{Sed}^{Dry} = 2.20_5 \times 10^3 kg\,m^{-3}$, which was obtained from the volume-fraction-weighted sums of the organic and mineral matter densities. The concentration of chemical in the dry sediment is now converted to $mol\,m^{-3}$ units as follows:

$$C_{Sed}^{Dry} = \frac{10.0\ \mu g\ g^{-1} \times 10^{-6}\ g\,\mu g^{-1} \times 10^3 g\,kg^{-1} \times 2.20_5 \times 10^3 kg\,m^{-3}}{354.5\,g\,mol^{-1}}$$

$$= 6.21_9 \times 10^{-2} mol\,m^{-3}$$

Since the sediment consists of only 0.20 solids by volume, the actual wet sediment concentration is:

$$C^0_{Sed} = 0.20 \times 6.21_9 \times 10^{-2} mol\, m^{-3} = 1.2_4 \times 10^{-2} mol\, m^{-3}$$

We have all the data we need to do the plots except for the maximum time step for stable integration. To determine this value, we need to decide which compartment has the shortest characteristic response time under the three scenarios. However, we must now include the effects of advection of the water column in our thinking. This is most easily done by assuming a constant background concentration of the chemical in the water column, which is constantly replenished as new water enters the region of interest. In this way, only the sediment fugacity will change with time, and we need only consider that when determining the maximum timestep.

For diffusion only, we have:

$$\tau_i = \frac{V_i Z_i}{D^L_i}$$

$$\tau^D_{Sed} = \frac{V_{Sed} Z^{Bulk}_{Sed}}{D^L_{Sed}} = \left(\frac{(1000\, m^2 \times 0.030\, m) \times 4.2_4 \times 10^3 Pa\, mol^{-1} m^{-3}}{4.1_7 \times 10^{-2} Pa\, mol^{-1} h^{-1}}\right) = 3.0_6 \times 10^6 h$$

For diffusion plus burial, we have:

$$\tau^{D+B}_{Sed} = \frac{V_{Sed} Z^{Bulk}_{Sed}}{D^L_{Sed}} = \left(\frac{(1000\, m^2 \times 0.030\, m) \times 4.2_4 \times 10^3 Pa\, mol^{-1} m^{-3}}{6.4_3 \times 10^{-1} Pa\, mol^{-1} h^{-1}}\right) = 1.9_8 \times 10^5 h$$

For diffusion, burial, and degradation, we have:

$$\tau^{D+B+Deg}_{Sed} = \frac{V_{Sed} Z^{Bulk}_{Sed}}{D^L_{Sed}} = \left(\frac{(1000\, m^2 \times 0.030\, m) \times 4.2_4 \times 10^3 Pa\, mol^{-1} m^{-3}}{1.3_2 Pa\, mol^{-1} h^{-1}}\right) = 9.6_3 \times 10^4 h$$

With conservative rounding to give round values for times, we find for diffusion alone (*D*), diffusion and burial (*D* + *B*), and all three processes including degradation (*D* + *B* + *Deg*):

$$\Delta t^D \leq 0.05 \times \tau^D_{Sed} = 1.5_3 \times 10^5 h \simeq 150{,}000\, h$$

$$\Delta t^{D+B} \leq 0.05 \times \tau^{D+B}_{Sed} = 9.9_0 \times 10^3 h \simeq 10{,}000\, h$$

$$\Delta t^{D+B+Deg} \leq 0.05 \times \tau^{D+B+Deg}_{Sed} = 4.8_2 \times 10^3 h \simeq 5{,}000\, h$$

With these results, we can see that the maximum timestep is greatest for diffusion-only losses as expected, since this is the process with the lowest *D*-value, i.e., the slowest process.

With the maximum time step in hand, we can set up the numerical integration. The shortest value is about 0.5 years, which makes use of the year as a time unit for plotting more sensible than hours, so we will include the necessary conversion from years to hours, i.e., multiply by 24 × 365.25 $h\ year^{-1}$. The change in sediment fugacity as a function of time is shown below for the three loss scenarios, with time steps of 15, 1, and 0.5 years used, respectively, for each scenario.

The initial fugacity in the sediment is simply the initial concentration divided by the bulk fugacity capacity:

$$f^0_{Sed} = \frac{C^0_{Sed}}{Z^{Bulk}_{Sed}} = \frac{1.2_4 \times 10^{-2} mol\, m^{-3}}{4.2_4 \times 10^3 Pa\, mol^{-1} m^{-3}} = 2.9_3 \times 10^{-6} Pa$$

For diffusion only, after 15 years the fugacity is given by the following expression, where $f_W^0 = C_W^0/Z_W^{Bulk}$:

$$f_{Sed}^{0+15} = f_{Sed}^0 + \left(\Delta t^{Diff} \times 24\,h\,day^{-1} \times 365.25\,day\,y^{-1}\right)\left(\frac{1}{V_{Sed} Z_{Sed}^{Bulk}}\right)\left(f_W^0 - f_{Sed}^0\right) D_{Sed-W}^{Diff-Ov}$$

For the change with diffusion and burial, and a time step of 1 year, the change in fugacity after 1 year is:

$$f_{Sed}^{0+1\,year} = f_{Sed}^0 + \left(\Delta t^{Diff+Burial} \times 24\,h\,day^{-1} \times 365.25\,day\,y^{-1}\right)\left(\frac{1}{V_{Sed} Z_{Sed}^{Bulk}}\right)\left(f_W^0 D_{Sed-W}^{Diff-Ov} - f_{Sed}^0 D_{Sed}^L\right)$$

Here, the total loss from sediment is due to both diffusive transfer to water and burial, the sum of the appropriate *D*-values being given by the following expression:

$$D_{Sed}^L = D_{Sed-W}^{Diff-Ov} + D_{Sed}^{Burial}$$

For the change with diffusion, burial, and degradation, and a time step of 0.5 years, the change in fugacity after 0.5 years is the same as for diffusion and burial alone, except for an additional degradation term in the total loss sum:

$$D_{Sed}^L = D_{Sed-W}^{Diff-Ov} + D_{Sed}^{Burial} + D_{Sed-W}^{Deg}$$

With these equations coded dynamically for each time step, the plots in Figure 6.15 were generated to demonstrate the time dependence of the sediment fugacity under the three scenarios (Figure 6.15).

The first 18 rows of the spread sheet used to generate this plot are shown in Figure 6.16.

Note that the time steps are different for each of the three scenarios. The number of steps to reach 50% loss ($t_{1/2}$) is about 15, and the number of steps to 95% loss is about 60, both of which are reasonable values for a numerical integration that is smooth and does not oscillate.

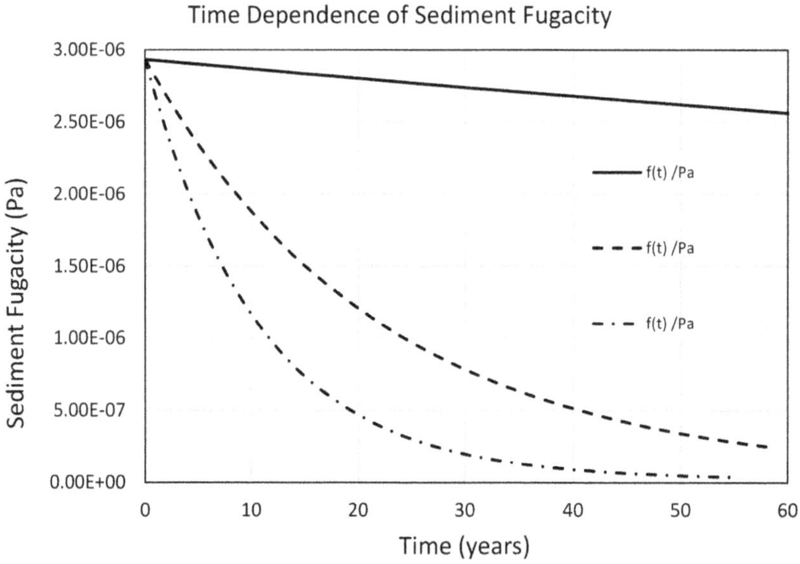

FIGURE 6.15 Three plots showing the time dependence of sediment fugacity under differing loss scenarios, namely diffusion-only (top), diffusion plus burial (middle), and diffusion, burial, and degradation (bottom).

	Diffusion Only			Diffusion, Burial			Diffusion, Burial, Degradation		
Step	Time (y)	f(t) (Pa)	f/f0	Time (y)	f(t) (Pa)	f/f0	Time (y)	f(t) (Pa)	f/f0
1	0.0	2.93E-06	1.00	0.0	2.93E-06	1.00	0.0	2.93E-06	1.00
2	15.0	2.83E-06	0.96	1.0	2.80E-06	0.96	0.5	2.80E-06	0.95
3	30.0	2.74E-06	0.92	2.0	2.68E-06	0.91	1.0	2.67E-06	0.91
4	45.0	2.65E-06	0.88	3.0	2.56E-06	0.87	1.5	2.55E-06	0.87
5	60.0	2.56E-06	0.84	4.0	2.45E-06	0.84	2.0	2.44E-06	0.83
6	75.0	2.48E-06	0.80	5.0	2.35E-06	0.80	2.5	2.33E-06	0.79
7	90.0	2.40E-06	0.77	6.0	2.24E-06	0.77	3.0	2.22E-06	0.76
8	105.0	2.33E-06	0.73	7.0	2.15E-06	0.73	3.5	2.12E-06	0.72
9	120.0	2.25E-06	0.70	8.0	2.05E-06	0.70	4.0	2.03E-06	0.69
10	135.0	2.18E-06	0.67	9.0	1.96E-06	0.67	4.5	1.93E-06	0.66
11	150.0	2.12E-06	0.64	10.0	1.88E-06	0.64	5.0	1.85E-06	0.63
12	165.0	2.05E-06	0.62	11.0	1.80E-06	0.61	5.5	1.76E-06	0.60
13	180.0	1.99E-06	0.59	12.0	1.72E-06	0.59	6.0	1.69E-06	0.57
14	195.0	1.94E-06	0.56	13.0	1.65E-06	0.56	6.5	1.61E-06	0.55
15	210.0	1.88E-06	0.54	14.0	1.57E-06	0.54	7.0	1.54E-06	0.52
16	225.0	1.83E-06	0.52	15.0	1.51E-06	0.51	7.5	1.47E-06	0.50
17	240.0	1.78E-06	0.49	16.0	1.44E-06	0.49	8.0	1.40E-06	0.48
18	255.0	1.73E-06	0.47	17.0	1.38E-06	0.47	8.5	1.34E-06	0.46

FIGURE 6.16 Screenshot of an Excel spreadsheet used for the calculation in Worked Example 6.6, showing the time, fugacity, and relative fugacity ratio for the three scenarios of diffusion only, diffusion and burial, and diffusion with burial and degradation.

6.8 SYSTEMS ACHIEVING EQUILIBRIUM

A slightly more complicated scenario than simple or parallel-process loss from a single compartment is one in which two or more compartments in contact achieve equilibrium following introduction of a chemical into one of them. In such a scenario, the concentrations, and therefore fugacities, change in all compartments.

Consider a simplified two-compartment system of water and sediment in which only diffusive exchange occurs. Under such an assumption, we ignore advection and degradation, as well as the many particle-based transfer processes associated with suspended particles and sediment resuspension and burial.

The differential rate expressions for the two compartments are as follows:

$$\frac{df_W}{dt} = \left(\frac{1}{V_W Z_W}\right)\left(D_{Sed-W}^{Diff-Ov} f_{Sed} - D_{Sed-W}^{Diff-Ov} f_W\right) = \left(\frac{D_{Sed-W}^{Diff-Ov}}{V_W Z_W}\right)\left(f_{Sed} - f_W\right)$$

$$\frac{df_{Sed}}{dt} = \left(\frac{1}{V_{Sed} Z_{Sed}}\right)\left(D_{Sed-W}^{Diff-Ov} f_W - D_{Sed-W}^{Diff-Ov} f_{Sed}\right) = \left(\frac{D_{Sed-W}^{Diff-Ov}}{V_{Sed} Z_{Sed}}\right)\left(f_W - f_{Sed}\right)$$

As introduced above, we can use the numerical integration approach to solve this problem. The outcome will depend on into which compartment the chemical is released, but in all cases, if the amount released is the same, the fugacities will be equal in all compartments at equilibrium, and the same partition ratios will be obtained. Worked Example 6.7 demonstrates these points.

Worked Example 6.7

Consider a system consisting of two compartments, A and B, with the following environmental and chemical parameters (based on the EQC model environment):

	Compartment A	**Compartment B**
V (m^3)	100	10.0
Z ($mol\ Pa^{-1}m^{-3}$)	2.00×10^{-2}	5.00×10^{-2}

100 moles of a chemical with the following properties is released into one of the compartments. Plot the changes in fugacity and concentration for release into compartment (i) A or (ii) B at time $t = 0$ until the system reaches equilibrium. Use the final concentrations to confirm the relevant partition ratio K_{AB}. Use an overall diffusion D-value of $D_{AB}^{Diff-Ov} = 1.52 \times 10^{-2}\ Pa\ mol^{-1}\ h^{-1}$.

(i) Release to compartment A:

As before, we set up the spreadsheet with the various parameters such as volumes, fugacity capacities, and rate constants. As well we have columns for time elapsed, the fugacity in each of the compartments, and their corresponding concentrations. To perform the calculation, we need first to determine the maximum time step value, based on the shortest characteristic time.

For the two compartments, we have:

$$\tau_A = \frac{1}{k_A} = \left(\frac{V_A Z_A}{D_{A-B}^{Diff-Ov}}\right) = \left(\frac{100\,m^3 \times 2.00 \times 10^{-2} Pa\,mol^{-1}m^{-3}}{1.52 \times 10^{-2} Pa\,mol^{-1}h^{-1}}\right) = 1.32 \times 10^{2} h$$

$$\tau_B = \frac{1}{k_B} = \left(\frac{V_B Z_B}{D_{A-B}^{Diff-Ov}}\right) = \left(\frac{10.0\,m^3 \times 5.00 \times 10^{-2} Pa\,mol^{-1}m^{-3}}{1.52 \times 10^{-2} Pa\,mol^{-1}h^{-1}}\right) = 3.29 \times 10^{1} h$$

Using the shortest time, τ_B, we can estimate the time step that is expected to lead to stable numerical integration as before:

$$\tau_{0.05} \leq 0.05 \times \tau_B = 0.05 \times \left(\frac{V_B Z_B}{D_{A-B}^{Diff-Ov}}\right) = 1.64\,h$$

For the sake of clarity, we will choose 1 h to keep the spreadsheet time in rounder units.

The spreadsheet setup used for the plot in Figure 6.18 is as shown in Figure 6.17a and b.

From this plot, we see that the system reaches equifugacity in about 150 h, which is our equilibrium criterion. The data reaches equifugacity after about 200 time steps, which is typical for a single equilibration process when the time step is chosen as 5% of the shortest response time.

The equilibrium concentrations are 8.0×10^{-1} and 2.0 $mol\ m^{-3}$ for compartments A and B. From this we can calculate the partition ratio K_{AB}:

$$K_{AB} = \frac{0.80}{2.0} = 0.40$$

From the data provided, we anticipate the following value from the ratio of fugacity capacities:

$$K_{AB} = \frac{Z_A}{Z_B} = \frac{0.020\,mol\,Pa^{-1}m^{-3}}{0.050\,mol\,Pa^{-1}m^{-3}} = 0.40$$

As is expected, these values agree exactly, as they must.

For comparison, the plots cast in terms of concentration for emission to compartment A are as shown in Figure 6.19.

(a)

	A	B	C	D	E
1	DelT (calc, h)	1.64	Dov(AB)	1.52E-02	mol Pa-1h-1
2	DelT (used,h)	1.00			
3					
4		V	Z	tau	Emission
5		m3	mol Pa-1m-3	h	moles
6	A	100	2.00E-02	131.58	100
7	B	10	5.00E-02	32.89	0
8					
9		Fugacity /Pa		Concentration /mol m-3	
10	Time /h	fA(t)	fB(t)	C(A)	C(B)
11	0	50.00	0.00	1.00	0.00
12	1	49.62	1.52	0.99	0.08
13	2	49.25	2.98	0.99	0.15
14	3	48.90	4.39	0.98	0.22
15	4	48.56	5.74	0.97	0.29
16	5	48.24	7.04	0.96	0.35

(b)

156	145	40.04	39.85	0.80	1.99
157	146	40.03	39.86	0.80	1.99
158	147	40.03	39.87	0.80	1.99
159	148	40.03	39.87	0.80	1.99
160	149	40.03	39.88	0.80	1.99
161	150	40.03	39.88	0.80	1.99
162	151	40.03	39.88	0.80	1.99
163	152	40.03	39.89	0.80	1.99
164	153	40.03	39.89	0.80	1.99
165	154	40.03	39.90	0.80	1.99

FIGURE 6.17 (a) and (b) Screenshots of an Excel spreadsheet for the calculation for chemical release to compartment *A* in Worked Example 6.7, showing the time, fugacities, and concentrations for compartments *A* and *B* at times 0–5 and 145–154 *h*. Also shown are the necessary input data used in the calculation.

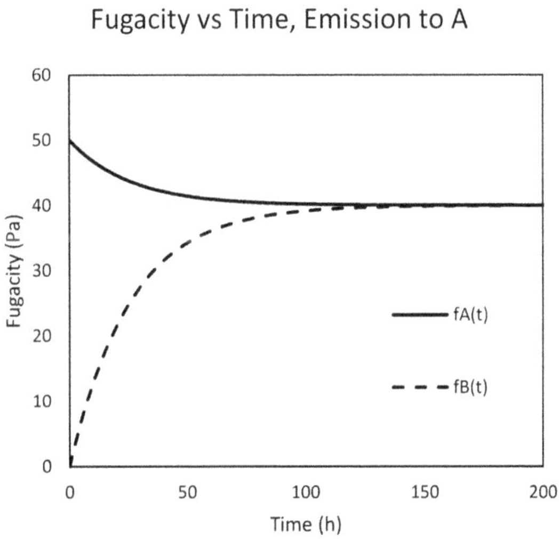

FIGURE 6.18 Plots showing the time dependence of the fugacity in two interacting compartments *A* and *B*, with initial chemical release to compartment *A*.

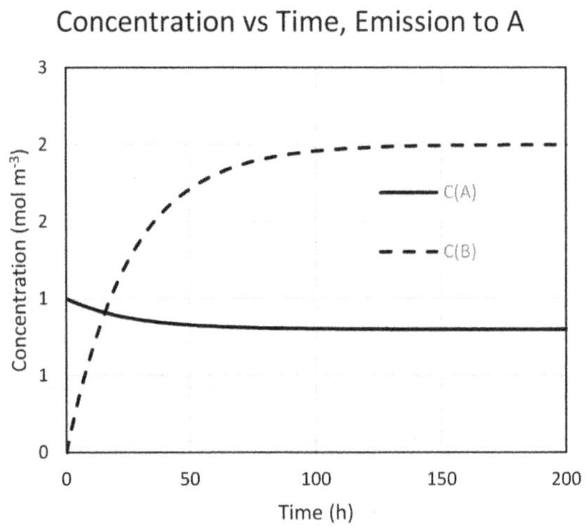

FIGURE 6.19 Plots showing the time dependence of concentration in two interacting compartments *A* and *B*, with initial chemical release to compartment *A*.

Viewed in this way, one can see that concentration itself is a less direct indicator of equilibrium, in that each curves reach a unique asymptotic limit, but these are not at the same value, unlike when the plots are cast in terms of fugacity. In cases where the various input data are not as convenient as this model situation, the concentration scale can vary over many orders of magnitude, making the use of a logarithmic scale necessary if all compartment concentrations are desired on the same graph.

(ii) Release to compartment *B*:
For release to compartment *B* instead, we have, by completely analogous calculation approaches, the calculation details shown in Figure 6.20 and the corresponding time dependence of fugacity shown in Figure 6.21.

	A	B	C	D	E
1	DelT (calc, h)	1.64	Dov(AB)	1.52E-02	mol Pa-1h-1
2	DelT (used,h)	1.00			
3					
4		V	Z	tau	Emission
5		m3	mol Pa-1m-3	h	moles
6	A	100	2.00E-02	131.58	0
7	B	10	5.00E-02	32.89	100
8					
9		Fugacity /Pa		Concentration /mol m-3	
10	Time /h	fA(t)	fB(t)	C(A)	C(B)
11	0	0.00	200.00	0.00	10.00
12	1	1.52	193.92	0.03	9.70
13	2	2.98	188.07	0.06	9.40
14	3	4.39	182.44	0.09	9.12
15	4	5.74	177.03	0.11	8.85
16	5	7.04	171.82	0.14	8.59

FIGURE 6.20 Screenshots of an Excel spreadsheet for the calculation in Worked Example 6.7, showing the time, fugacities, and concentrations for compartments *A* and *B* at times 0–5 *h* for chemical release to compartment *B*. Also shown are the necessary input data used in the calculation.

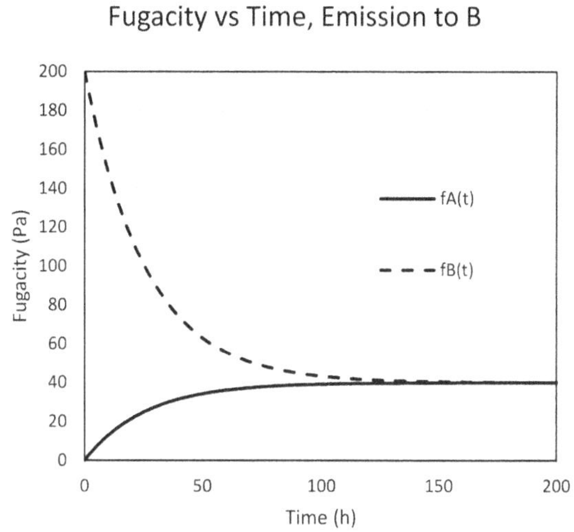

FIGURE 6.21 Plots showing the time dependence of the fugacity in two interacting compartments *A* and *B*, with initial chemical release to compartment *B*.

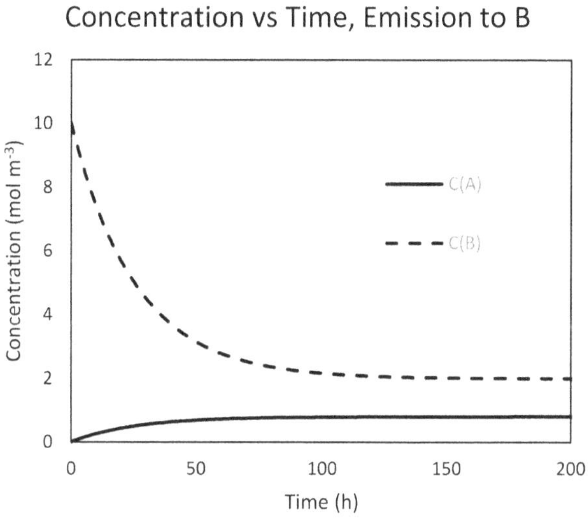

FIGURE 6.22 Plots showing the time dependence of concentration in two interacting compartments *A* and *B*, with initial chemical release to compartment *B*.

Note that the same fugacity is achieved at equilibrium, irrespective of where the chemical is released.

Finally, the same results plotted in terms of concentration (Figure 6.22), yields the same final concentrations for release in either compartment.

It is evident that the same partition ratio, $K_{AB} = 0.80/2.0 = 0.40$, is obtained, as anticipated for the same amount of chemical equilibrating within the same system.

6.9 MORE COMPLEX KINETIC SITUATIONS

As the number of compartments increases, along with the number of transfer and loss mechanisms, the modelling challenge grows significantly. As a demonstration of such a system, consider the Level III model which includes air, water, soil, and sediment, with degradation at different rates in each compartment and advection in all compartments but soil. Intercompartment transfer mechanisms include air-to-water transport by dry and wet deposition of aerosols, water-to-sediment deposition of suspended particles, losses from sediment through resuspension and burial, and transfer of soil water and solids to water through runoff, and other processes. With all these coupled processes, we arrive at a rather complex situation when crafting the differential rate expressions. However, the numerical integration approach is the same, albeit with many more terms in the equations!

For example, take the Level III environment system developed in Section 5.3. The complete list of D-values for the processes considered in this model is given below.

Process	Symbol	Process	Symbol
Air–water diffusion	$D_{AW}^{Diff-Ov}$	Air advection	D_{A}^{Adv}
Air–soil diffusion	$D_{A-Soil}^{Diff-Ov}$	Water advection	D_{W}^{Adv}
Water–sediment diffusion	$D_{Sed-W}^{Diff-Ov}$	Sediment burial	D_{Sed}^{Burial}
Rain (air–water)	D_{AW}^{Rain}	Air degradation	D_{A}^{Deg}
Aerosol deposition (air–water)	D_{AW}^{Q}	Water degradation	D_{W}^{Deg}
Rain (air–soil)	D_{A-Soil}^{Rain}	Soil degradation	D_{Soil}^{Deg}
Aerosol deposition (air–soil)	D_{A-Soil}^{Q}	Sediment degradation	D_{Sed}^{Deg}
Sediment deposition	D_{Sed-W}^{Dep}	Soil water runoff	$D_{Soil-W}^{Runoff-W}$
Sediment resuspension	D_{Sed-W}^{Resusp}	Soil organic matter runoff	$D_{Soil-W}^{Runoff-Solids}$

Assuming there is no emission to any compartment after time $t = 0$, the corresponding differential equations for each of the four main compartments follow, constructed in the manner introduced conceptually in Sections 6.1 and 6.2:

$$\frac{df_A}{dt} = \left(\frac{1}{V_A Z_A}\right)\left(\begin{array}{l} \left[D_{AW}^{Diff-Ov} f_W + D_{A-Soil}^{Diff-Ov} f_{Soil} + D_{A}^{Adv} f_A^{In}\right] \\ -\left[\left(D_{AW}^{Diff-Ov} + D_{AW}^{Rain} + D_{AW}^{Q}\right) + \left(D_{A-Soil}^{Diff-Ov} + D_{A-Soil}^{Rain} + D_{A-Soil}^{Q}\right) + D_{A}^{Adv} + D_{A}^{Deg}\right] f_A \end{array}\right)$$

$$\frac{df_W}{dt} = \left(\frac{1}{V_W Z_W}\right)\left(\begin{array}{l} \left[\begin{array}{l} \left(D_{AW}^{Diff-Ov} + D_{AW}^{Rain} + D_{AW}^{Q}\right) f_A + \left(D_{Soil-W}^{Runoff-W} + D_{Soil-W}^{Runoff-Solids}\right) f_{Soil} \\ +\left(D_{Sed-W}^{Diff-Ov} + D_{Sed-W}^{Resusp}\right) f_{Sed} + D_{W}^{Adv} f_W^{In} \end{array}\right] \\ -\left[D_{AW}^{Diff-Ov} + \left(D_{Sed-W}^{Diff-Ov} + D_{Sed-W}^{Dep}\right) + D_{W}^{Adv} + D_{W}^{Deg}\right] f_W \end{array}\right)$$

$$\frac{df_{Soil}}{dt} = \left(\frac{1}{V_{Soil} Z_{Soil}}\right)\left(\begin{array}{l} \left[\left(D_{A-Soil}^{Diff-Ov} + D_{A-Soil}^{Rain} + D_{A-Soil}^{Q}\right) f_A\right] \\ -\left[D_{A-Soil}^{Diff-Ov} + D_{Soil-W}^{Runoff-W} + D_{Soil-W}^{Runoff-Solids} + D_{Soil}^{Deg}\right] f_{Soil} \end{array}\right)$$

$$\frac{df_{Sed}}{dt} = \left(\frac{1}{V_{Sed} Z_{Sed}}\right)\left(\begin{array}{l} \left[\left(D_{Sed-W}^{Diff-Ov} + D_{Sed-W}^{Dep}\right) f_W\right] \\ -\left[D_{Sed-W}^{Diff-Ov} + D_{Sed-W}^{Resusp} + D_{Sed}^{Burial} + D_{Sed}^{Deg}\right] f_{Sed} \end{array}\right)$$

In each equation, the following principles are true:

1. Volumes and fugacity capacities correspond to the compartment in question.
2. All positive terms are due to input processes that move chemical *into* the compartment in question.

3. All negative terms are due to output processes that move chemical *out of* the compartment in question.

The differential rate laws for each compartment may be more compactly written out by representing the sum of all D-values for losses from each compartment with a single symbol D_i^L as follows:

$$D_A^L = \left(D_{AW}^{Diff-Ov} + D_{AW}^{Rain} + D_{AW}^{Q}\right) + \left(D_{A-Soil}^{Diff-Ov} + D_{A-Soil}^{Rain} + D_{A-Soil}^{Q}\right) + D_A^{Adv} + D_A^{Deg}$$

$$D_W^L = D_{AW}^{Diff-Ov} + \left(D_{Sed-W}^{Diff-Ov} + D_{Sed-W}^{Dep}\right) + D_W^{Adv} + D_W^{Deg}$$

$$D_{Soil}^L = D_{A-Soil}^{Diff-Ov} + \left(D_{Soil-W}^{Runoff-W} + D_{Soil-W}^{Runoff-Solids}\right) + D_{Soil}^{Deg}$$

$$D_{Sed}^L = \left(D_{Sed-W}^{Diff-Ov} + D_{Sed-W}^{Resusp}\right) + D_{Sed}^{Burial} + D_{Sed}^{Deg}$$

Here, parallel transfer processes between two compartments are grouped with parentheses for clarity.

With these definitions, we can now write the differential rate balances in their more compact form as follows:

$$\frac{df_A}{dt} = \left(\frac{1}{V_A Z_A}\right)\left[D_{AW}^{Diff-Ov} f_W + D_{A-Soil}^{Diff-Ov} f_{Soil} + D_A^{Adv} f_A^{In} - D_A^L f_A\right]$$

$$\frac{df_W}{dt} = \left(\frac{1}{V_W Z_W}\right)\left[\begin{array}{l}\left(D_{AW}^{Diff-Ov} + D_{AW}^{Rain} + D_{AW}^{Q}\right) f_A + \left(D_{Soil-W}^{Runoff-W} + D_{Soil-W}^{Runoff-Solids}\right) f_{Soil} \\ + \left(D_{Sed-W}^{Diff-Ov} + D_{Sed-W}^{Resusp}\right) f_{Sed} + D_W^{Adv} f_W^{In} - D_W^L f_W\end{array}\right]$$

$$\frac{df_{Soil}}{dt} = \left(\frac{1}{V_{Soil} Z_{Soil}}\right)\left[\left(D_{A-Soil}^{Diff-Ov} + D_{A-Soil}^{Rain} + D_{A-Soil}^{Q}\right) f_A - D_{Soil}^L f_{Soil}\right]$$

$$\frac{df_{Sed}}{dt} = \left(\frac{1}{V_{Sed} Z_{Sed}}\right)\left[\left(D_{Sed-W}^{Diff-Ov} + D_{Sed-W}^{Dep}\right) f_W - D_{Sed}^L f_{Sed}\right]$$

We may now use these equations to explore some situations in the EQC environment, and their corresponding time responses of fugacity and concentration. Such calculations are referred to as Level IV in the Mackay fugacity-based modelling nomenclature.

Worked Example 6.8

Use the differential equations introduced for the EQC environment above to model the time-dependence of the redistribution to equilibrium of 1000 *kg* of a DDT-like chemical (molar mass 354.5 *g mol*$^{-1}$) released to each of air, water, soil, or sediment, in turn. Assume inflowing air and water by advection is free of DDT and use the following data for your simulations:

Z_A^{Bulk}	4.034×10^{-4} *mol Pa*$^{-1}$ *m*$^{-3}$	V_A	1.0×10^{14} *m*^{3}
Z_W^{Bulk}	7.757×10^{-1} *mol Pa*$^{-1}$ *m*$^{-3}$	V_W	2.0×10^{11} *m*^{3}
Z_{Soil}^{Bulk}	1.0090×10^{4} *mol Pa*$^{-1}$ *m*$^{-3}$	V_{Soil}	1.8×10^{10} *m*^{3}
Z_{Sed}^{Bulk}	8.0750×10^{3} *mol Pa*$^{-1}$ *m*$^{-3}$	V_{Sed}	5.0×10^{8} *m*^{3}

D-values (*mol Pa*⁻¹ *h*⁻¹):

$D_{AW}^{Diff-Ov}$	1.92×10^7	$D_{Soil-W}^{Runoff-W}$	3.49×10^6	D_{W}^{Adv}	2.99×10^8
D_{AW}^{Rain}	7.76×10^5	$D_{Soil-W}^{Runoff-Solids}$	1.82×10^7	D_{Sed}^{Burial}	1.00×10^4
D_{AW}^{Q}	1.08×10^8	$D_{Sed-W}^{Diff-Ov}$	7.76×10^5	D_{A}^{Deg}	2.49×10^8
$D_{A-Soil}^{Diff-Ov}$	1.41×10^6	D_{Sed-W}^{Dep}	6.31×10^8	D_{W}^{Deg}	7.43×10^6
D_{A-Soil}^{Rain}	6.98×10^6	D_{Sed-W}^{Resusp}	8.07×10^7	D_{Soil}^{Deg}	4.35×10^9
D_{A-Soil}^{Q}	9.76×10^8	D_{A}^{Adv}	7.65×10^8	D_{Sed}^{Deg}	2.15×10^7

Total loss *D*-value sums (*mol Pa*⁻¹ *h*⁻¹):

D_{A}^{L}	2.13×10^9	D_{Soil}^{L}	4.37×10^9
D_{W}^{L}	9.58×10^8	D_{Sed}^{L}	1.03×10^8

The spreadsheet used to generate the plots in Figures 6.24–6.27 for this problem is shown in Figure 6.23. The sheet comprises entries for all the necessary *D*-values, *Z*-values, and other input data, as well as the calculated dependence of each compartment's fugacity and concentration on time. The equations used to calculate all but the first row of fugacities and concentrations are as follows, as exemplified by those of the second time step row at $t = 1\ h$:

Air fugacity:

$$G14 = G13 + \$H\$2 * \left(1/(\$H\$6 * \$I\$6)\right) * (\$B\$3 * H13 + \$B\$6 * I13 + \$B\$14 * \$N\$6 - \$B\$22 * G13)$$

Water fugacity:

$$H14 = H13 + \$H\$2 * \left(1/(\$H\$7 * \$I\$7)\right) * \begin{pmatrix} (\$B\$3 + \$B\$4 + \$B\$5) * G13 + (\$B\$9 + \$B\$10) * I13 \\ +(\$B\$11 + \$B\$13) * J13 + \$B\$15 * \$N\$7 - \$B\$23 * H13 \end{pmatrix}$$

	A	B	C	D	E	F	G	H	I	J	K	L	M	N
1	D-values:			Z-values			DelT (calc, h)	9.48E-01		MM (kg mol-1)		3.55E-01		
2							DelT (used, h)	1.00E+00						
3	DovAW	1.92E+07		Airbulk	0.000									
4	DrainAW	7.76E+05		Wbulk	0.776			V	Z	tau	Emission	Emission	C_0	C bg
5	DqAW	1.08E+08		Soilbulk	10100.000			m^3	mol Pa^{-1} m^{-3}	h	kg	mol	mol m^{-3}	mol m^{-3}
6	DovASoil	1.41E+06		Sedbulk	8080.000		Air	1.00E+14	4.03E-04	1.90E+01	1.00E+03	2.82E+03	2.82E-11	0
7	DrainASoil	6.98E+06					Water	2.00E+11	7.76E-01	1.62E+02	0.00E+00	0	0.00E+00	0
8	DqASoil	9.76E+08					Soil	1.80E+10	1.01E+04	4.16E+04	0.00E+00	0	0.00E+00	0
9	DrunW	3.49E+06					Sediment	5.00E+08	8.08E+03	3.92E+04	0.00E+00	0	0.00E+00	0
10	DrunOM	1.82E+07												
11	DovSedW	7.76E+05			Time		Fugacity /Pa							
12	DdepSedW	6.31E+08			weeks	hours	fA(t)	fW(t)	f(Soil)	f(Sed)				
13	DresSedW	8.07E+07			0.000	0.0	7.00E-08	0.00E+00	0.00E+00	0.00E+00				
14	DadvA	7.65E+08			0.006	1.0	6.63E-08	5.77E-11	3.79E-13	0.00E+00				
15	DadvW	2.99E+08			0.012	2.0	6.28E-08	1.12E-10	7.38E-13	9.03E-15				
16	DadvSe	1.00E+04			0.018	3.0	5.95E-08	1.63E-10	1.08E-12	2.65E-14				
17	DdegA	2.49E+08			0.024	4.0	5.64E-08	2.11E-10	1.40E-12	5.21E-14				
18	DdegW	7.43E+06			0.030	5.0	5.34E-08	2.56E-10	1.71E-12	8.51E-14				
19	DdegSo	4.35E+09			0.036	6.0	5.06E-08	2.99E-10	1.99E-12	1.25E-13				
20	DdegSe	2.15E+07			0.042	7.0	4.79E-08	3.39E-10	2.27E-12	1.72E-13				
21					0.048	8.0	4.54E-08	3.76E-10	2.53E-12	2.25E-13				
22	DlossA	2.13E+09			0.054	9.0	4.30E-08	4.11E-10	2.77E-12	2.84E-13				
23	DlossW	9.58E+08			0.060	10.0	4.07E-08	4.44E-10	3.01E-12	3.48E-13				
24	DlossSo	4.37E+09			0.065	11.0	3.86E-08	4.75E-10	3.23E-12	4.17E-13				
25	DlossSe	1.03E+08			0.071	12.0	3.65E-08	5.04E-10	3.43E-12	4.92E-13				
26					0.077	13.0	3.46E-08	5.31E-10	3.63E-12	5.70E-13				

FIGURE 6.23 Screenshot of the spreadsheet used to generate the plots for Worked Example 6.8, with the results for the first 12 *h* shown.

Soil fugacity:

$$I14 = I13 + \$H\$2 * (1/(\$H\$8 * \$I\$8)) * ((\$B\$6 + \$B\$7 + \$B\$8) * G13 - \$B\$24 * I13)$$

Sediment fugacity:

$$J14 = J13 + \$H\$2 * (1/(\$H\$9 * \$I\$9)) * ((\$B\$11 + \$B\$12) * H13 - \$B\$25 * J13)$$

Each expression takes the fugacity for the compartment in question at the prior time step (one cell above) and adds to it the change that occurs over the small time step. This change is calculated as the time step multiplied by the reciprocal product of the compartment volume and fugacity capacity, itself multiplied by the sum of all *D*-values for inputs to the compartment minus the sum of all losses from the compartment in question, each multiplied by their respective fugacities at the previous time step. The same equations are carried down for as many cells as are needed to complete the desired time of simulation, in this case, about 1700 cells with the conservative time step chosen.

Using such a spreadsheet approach, the plots of the type given in Figures 6.24–6.27 may be generated. These plots demonstrate the variation in chemical fate and distribution behaviour within a complex environmental system involving many coupled compartments, each with varying efficiencies of intercompartment transfer and differing rates of degradation and flushing by advection. Each set of curves reflects the point of emission and the relative efficiency of intercompartment transfer.

Release to air has relatively efficient mechanisms associated with transfer to water and soil, but not to sediment. Hence, the sediment fugacity, which is fed only from transfer from water, is delayed both by the time for water to build up a significant fugacity, and by the relatively inefficient diffusive and depositional transfer from water, its only chemical source. The water fugacity itself initially grows to a point of near-equilibrium with the air, but then decreases over a much longer time due to transfer to the sediment, which has a much greater fugacity capacity for this chemical. In contrast, the soil fugacity builds up due to transfer from the air, but since the rate of runoff transfer is relatively slow, it reaches a steady-state value early and more or less remains at that value over the timescale of the simulation.

Release to water shows rapid achievement of a pseudo-steady state between the air and water, followed by much slower transfer to soil, due to wet and dry deposition, and to sediment through relatively slow water–sediment transfer.

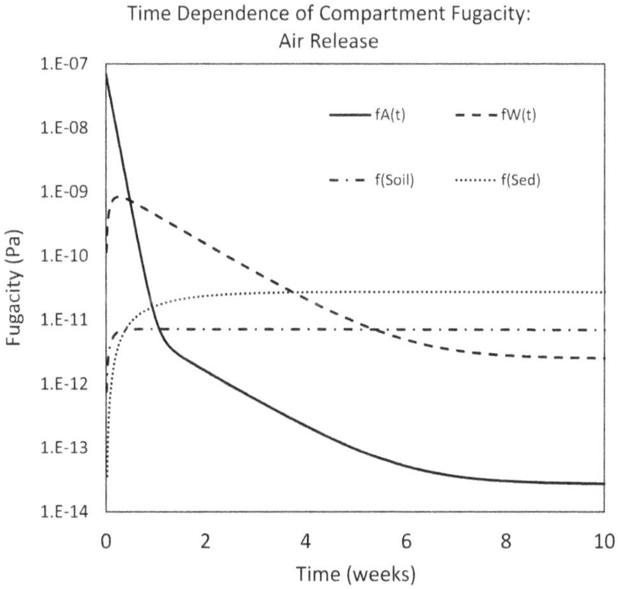

FIGURE 6.24 Plots showing the time dependence of fugacity in air, water, soil, and sediment compartments, with initial chemical release to air.

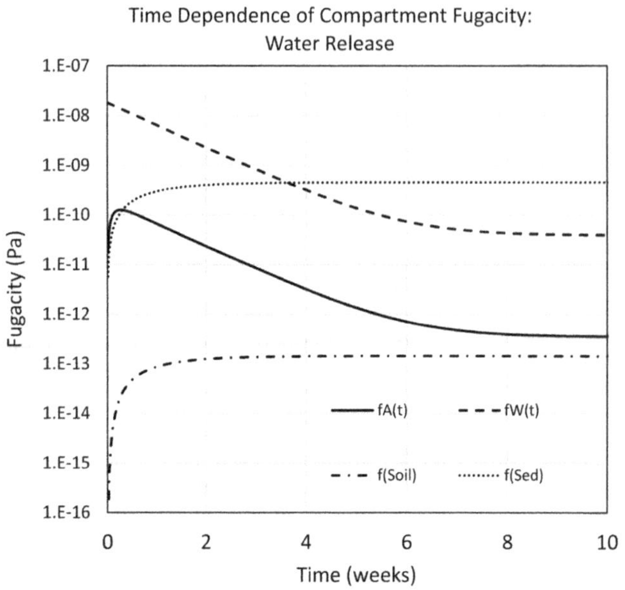

FIGURE 6.25 Plots showing the time dependence of fugacity in air, water, soil, and sediment compartments, with initial chemical release to water.

Soil release gives a very different picture, since the high hydrophobicity of the chemical makes soil a good chemical sink due to its large fugacity capacity. Transfer to air is relatively fast but does not significantly raise the fugacity due to the constant advective "cleaning" of the air. Soil–water transfer is slower, relying only on soil–water runoff which itself is slow and also competes with advective losses in the water. Finally, the sediment uptake is slowest, mostly due to the combination of the limited runoff transfer rate and the slow water-sediment transfer processes that follow.

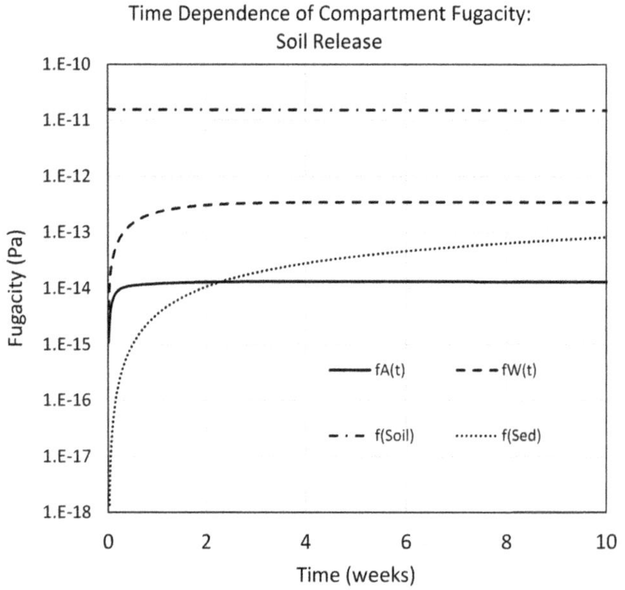

FIGURE 6.26 Plots showing the time dependence of fugacity in air, water, soil, and sediment compartments, with initial chemical release to soil.

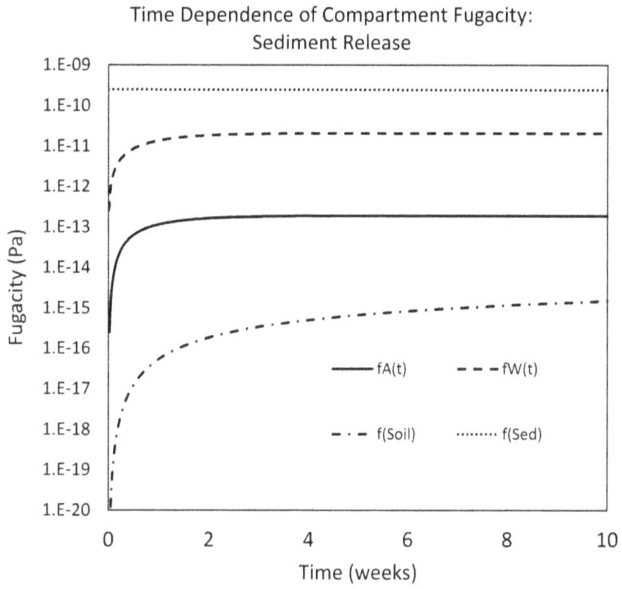

FIGURE 6.27 Plots showing the time dependence of fugacity in air, water, soil, and sediment compartments, with initial chemical release to sediment.

Finally, direct release into the sediment gives essentially immediate steady-state conditions in that compartment due to the high fugacity capacity of sediment and low burial and resuspension rates. All other compartments achieve a steady-state condition in the following order: water which directly contacts the sediment; air, which achieves transfer indirectly via water; and soil which required deposition and diffusive transfer from the air, the slowest chain of transfer.

6.10 CONCLUDING REMARKS

At this stage, modelling work on environmental systems becomes a many-faceted elaboration of the concepts built up to this point. Specific modelling scenarios will require consideration of the unique properties of both the chemical(s) in question and the environment within which they move. As well, extensions into biological and physiological systems bring new concepts and approaches that are beyond the scope of this workbook. Nevertheless, the ideas and basic concepts remain the same. For environmental systems, a chemical will always move toward lower fugacity until equilibrium is achieved or the system changes in some way.

Symbols Used in This Book

Chapter 1: Equilibrium Partitioning in Closed Systems: All About *Z*-Values

Symbol	Name	Unit	Comment
C_i	Concentration in medium "*i*"	$mol\ m^{-3}$	
m	Moles of matter in system	mol	
V_i	Volume of compartment "*i*"	m^3	
M	Molar mass of chemical species	$g\ mol^{-1}$	
K_{ij}	Partition ratio between compartments "*i*" and "*j*"	*unitless*	Some published partition ratios have units, depending on how the ratio is defined.
m_i	Moles of matter in compartment "*i*"	mol	
Z_i	Fugacity capacity of medium "*i*" for a given chemical	$mol\ Pa^{-1}\ m^{-3}$	Every chemical has its own unique value for any medium.
f_i	Fugacity of a given chemical in medium "*i*"	Pa	
f_{Sys}	System fugacity for an equilibrium system	Pa	Does not exist for systems that are not at equilibrium.
P_i	Partial pressure of chemical "*i*"	Pa	
R	Gas constant	$m^3\ Pa\ K^{-1}\ mol^{-1}$	There are many other possible units and values
T	Temperature	K	Beware temperatures quoted in Celsius which require conversion
H	Henry's Law constant	$Pa\ m^3\ mol^{-1}$	As with the gas constant, there are many other units used.
C_i^{Sat}	Saturation concentration in medium "*i*"	$mol\ m^{-3}$	
P_A^{Sat}	Partial pressure in air of a saturated solution	Pa	
$P_A^{Sat-SCL}$	Theoretical subcooled liquid vapour pressure in air of a saturated solution	Pa	
F	Fugacity ratio	*unitless*	
x_i	Mole fraction of a chemical in medium "*i*"	*unitless*	
v_i^{f-j}	Volume fraction of "*j*" in medium "*i*".	*unitless*	
ρ_i	Density of medium "*i*"	$kg\ m^{-3}$	
m_i^{f-j}	Mass fraction of "*j*" in medium "*i*"	*unitless*	

Chapter 2: Equilibrium Partitioning in Complex Media: Bulk *Z*-Values

Symbol	Name	Unit	Comment
Z_i^{Bulk}	Bulk fugacity capacity of a chemical in medium "*i*"	$mol\ Pa^{-1}\ m^{-3}$	A volume-fraction-weighted sum of component fugacity capacities.
C_i^{Bulk}	Bulk concentration of a chemical in medium "*i*"	$mol\ m^{-3}$	
K_P	Aerosol–air partition ratio	$m^3\ \mu g^{-1}$	Mass-based unit mass of chemical per μg aerosol over mass chemical per m^3 air.

Chapter 3: Open Systems at Steady State: Introducing *D*-values

Symbol	Name	Unit	Comment
r_i^j	Molar flow rate of chemical in medium "*i*" by process "*j*"	$mol\ h^{-1}$	
G_i^j	Flow rate of medium "*i*" by process "*j*"	$m^3\ h^{-1}$	
E	Emission rate to system	$mol\ h^{-1}$	May also be in mass per time units such as $kg\ h^{-1}$.
I	Total molar input rate for a chemical	$mol\ h^{-1}$	
D_i^{Adv}	*D*-value for advective flow associated with medium "*i*"	$mol\ Pa^{-1}\ h^{-1}$	
D_i^{Deg}	*D*-value for degradation loss associated with medium "*i*"	$mol\ Pa^{-1}\ h^{-1}$	
V_i	Volume of a medium "*i*"	m^3	
U^j	Velocity of a process "*j*"	$m\ h^{-1}$	Often this unit is arrived at indirectly and may not be of obvious origin.
A_i	Surface area of an interface of medium "*i*"	m^2	
Q	Scavenging ratio	unitless	Volume of air scavenged per volume of aerosol deposited.
L_i^j	Mass transfer flux into medium "*i*" by process "*j*" across an interfacial area per time	$kg\ m^{-2}\ h^{-1}$	
k_i^{Deg}	Degradation rate constant for chemical in medium "*i*"	h^{-1}	Generally assumed to be a first-order or pseudo-first-order process for environmental contaminants.
r^L	Total loss rate	$mol\ h^{-1}$	
D_i^L	Total *D*-value sum for all loss processes from medium "*i*"	$mol\ Pa^{-1}\ h^{-1}$	
τ_i^j	Residence time of chemical in medium "*i*" due to loss by process "*j*"	h	May be in more convenient time units such as days, months, years, depending on context.
τ^L	Residence time of chemical due to all loss processes combined	h	May be in more convenient time units such as days, months, years, depending on context.

Chapter 4: Non-Equilibrated Open Systems: *D*-values for Diffusive Transport

Symbol	Name	Unit	Comment
k_i^M	Mass transfer coefficient associated with medium "*i*"	$m\ h^{-1}$	
B_i	Molecular diffusivity of a chemical in medium "*i*"	$m^2\ h^{-1}$	
Y	Diffusion distance	m	
$D_{ij}^{Diff-Ov}$	Overall diffusion *D*-value between media "*i*" and "*j*"	$mol\ Pa^{-1}\ h^{-1}$	Since net diffusion may be in either direction, the subscript indices are given in alphabetical order for consistency.
k_{ij}^{M-i}	"*i*"-side mass-transfer coefficient for interface between media "*i*" and "*j*"	$m\ h^{-1}$	
D_{ij}^T	Total *D*-value for all intermedia transfer processes between media "*i*" and "*j*"	$mol\ Pa^{-1}\ h^{-1}$	
B_j^{Eff-i}	Effective diffusivity in component "*i*" of medium "*j*"	$m^2\ h^{-1}$	
E_i	Emission rate to medium "*i*"	$mol\ h^{-1}$	

Chapter 5: Basic Environmental Models: Putting It All Together

Symbol	Name	Unit	Comment
$\tau_i^{1/2-j}$	Half-time for process or subcompartment "*j*" in medium "*i*"	h	May be in other more convenient units such as days, months, years, etc.

Chapter 6: Time-Variant Systems: Differential Equations

Symbol	Name	Unit	Comment
Δt	Time step for numerical integration	h	This is a recommendation and can be "played with".

References

Harner, T.; Bidleman, T.F. (1998) "Octanol–Air Partition Coefficient for Describing Particle/Gas Partitioning of Aromatic Compounds in Urban Air". *Environ. Sci. Technol*, *32*, 1494–1502.

Hughes, L.; Mackay, D.; Powell, D.E.; Kim, J. (2012) "An Updated State of the Science EQC Model for Evaluating Chemical Fate in the Environment: Application to D5 (decamethylcyclopentasiloxanne)". *Chemosphere*, 2012, *87*, 118–124.

Jury, W.A.; Spencer, W.F.; Farmer, W.J. (1983) "Use of Models for Assessing Relative Volatility, Mobility, and Persistence of Pesticides and Other Trace Organics in Soil Systems" in *Hazard Assessment of Chemicals*. J. Saxena, Ed. Academic Press.

Karickhoff, S.W. (1981) "Semi-Empirical Estimation of Sorption of Hydrophobic Pollutants on Natural Sediments and Soils". *Chemosphere*, *10*, 833–846.

Mackay, D.; Paterson, S.; Shui, W.Y. (1992) "Generic Models for Evaluating the Regional Fate of Chemicals". *Chemosphere*, *24*, 695–717.

Mackay, D.; Peterson, S.; Shroeder, W.H. (1986) "Model Describing the Rates of Transfer Processes of Organic Chemicals between Atmosphere and Water". *Environ. Sci. Technol*, *20*, 810–816.

Parnis, J.M.; Mackay, D. (2021) "*Multimedia Environmental Models: The Fugacity Approach*" 3rd edition CRC Press, Boca Raton FL.

Seth, R.; Mackay, D.; Munck, A. (1999) "Estimating the Organic Carbon Partition Coefficient and Its Variability for Hydrophobic Chemicals". *Environ. Sci. Technol.*, *33*, 2390–2394.

Sposito, G. (1989) "*The Chemistry of Soils*" Oxford University Press, New York.

Index

E

F

H

I

K

L

M

N

W

Z

For Product Safety Concerns and Information please contact our EU
representative GPSR@taylorandfrancis.com
Taylor & Francis Verlag GmbH, Kaufingerstraße 24, 80331 München, Germany

www.ingramcontent.com/pod-product-compliance
Lightning Source LLC
Chambersburg PA
CBHW081530220326
41598CB00036B/6383
9781041108658